Q175 THA

The Cognitive Science of Science

The Cognitive Science of Science: Explanation, Discovery, and Conceptual Change

Paul Thagard

in collaboration with Scott Findlay, Abninder Litt, Daniel Saunders, Terrence C. Stewart, and Jing Zhu

The MIT Press

Cambridge, Massachusetts

London, England

For information about special quantity discounts, please email special_sales@mitpress.mit.edu

This book was set in Stone Sans and Stone Serif by Toppan Best-set Premedia Limited. Printed and bound in the United States of America.

Library of Congress Cataloging-in-Publication Data

Thagard, Paul.
The cognitive science of science : explanation, discovery, and conceptual change / Paul Thagard ; in collaboration with Scott Findlay . . . [et al.].
    p. cm.
Includes bibliographical references and index.
ISBN 978-0-262-01728-2 (hardcover : alk. paper)
1. Science—Philosophy. 2. Cognitive science. I. Findlay, Scott. II. Title.
Q175.T478   2012
501—dc23

                                                                    2011039760

10 9 8 7 6 5 4 3 2 1

To the pioneers of the cognitive science of science, especially Herbert Simon

Contents

Preface

This book is a collection of my recent essays on the cognitive science of science that illustrate ways of combining philosophical, historical, psychological, computational, and neuroscientific approaches to explaining scientific development. Most of the chapters have been or will be published elsewhere, but the introductions are brand new (chapters 1, 2, 7, 12), as are the last two chapters, which take the cognitive science of science in new directions related to values and concepts. The reprinted chapters reproduce the relevant articles largely intact, but I have done some light editing to coordinate references and remove redundancies. Origins of the articles and coauthors are indicated in the acknowledgments.

Early in my career, I wandered into the cognitive science of science through a series of educational accidents, and have enthusiastically pursued research that is variously philosophical, historical, psychological, computational, and neurobiological. In high school, I did very well in physics and chemistry, but only because I was adept at solving math problems, not because I found science very interesting. As an undergraduate at the University of Saskatchewan, I avoided serious science courses, although I did get a good sampling of mathematics and logic. My interest in science was sparked during my second undergraduate degree at Cambridge University, where the philosophy course I took required a paper in philosophy of science. Through lectures by Ian Hacking and Gerd Buchdahl, along with books by Russell Hanson and Thomas Kuhn, I started to appreciate the value of understanding the nature of knowledge by attention to the history of science. I was struck by how much more rich and interesting the scientific examples of knowledge were compared to the contrived thought experiments favored by epistemologists working in the tradition of analytic philosophy. Accordingly, my Ph.D. work at the

University of Toronto focused on scientific reasoning enriched by historical case studies.

My move into cognitive science was also serendipitous. In 1978, in the second term of my first teaching job at the University of Michigan–Dearborn, I decided to sit in on a graduate epistemology course taught by Alvin Goldman at the main Michigan campus in Ann Arbor. It turned out that this course was coordinated with one on human inference taught by the social psychologist Richard Nisbett. The combination of these two courses was amazing: Goldman was pioneering an approach to epistemology that took experimental research on psychology seriously, and Nisbett was presenting a draft of his path-breaking book with Lee Ross, *Human Inference.* I started reading avidly in cognitive psychology, which quickly led me to the field of artificial intelligence. I was attracted by the theoretical ideas of visionaries such as Marvin Minsky, and also by the prospects of a new methodology—computer modeling—for understanding the structure and growth of scientific knowledge.

Accordingly, I did an MS in computer science at Michigan and started building my own computational models of various aspects of scientific thinking. My early models of analogical thinking were somewhat crude, but became much more powerful when my collaborator Keith Holyoak came up with the idea of modeling analogy using connectionist ideas about parallel constraint satisfaction. I quickly realized that theory choice based on the explanatory power of competing theories could also be simulated using neural networks.

My interest in neuroscience also came about indirectly. After I moved to Waterloo in 1992, one of my graduate students, Allison Barnes, was investigating empathy as a kind of analogy, which led me to general concern with emotions. The work of Antonio Damasio revealed how crucial neuroscience was to understanding emotions, and since I was already building artificial neural network models, it was natural to try to undertake more realistic neural models of emotion and decision making. Happily, this line of work has turned back around to scientific applications, described in some of the chapters below.

I remain convinced that understanding the growth of knowledge requires the kind of interdisciplinary approach found in cognitive science. I hope this collection will appeal to anyone interested in the structure and growth of scientific knowledge, including scientists, philosophers, historians, psychologists, sociologists, and educators.

Acknowledgments

While I was writing and revising this work, my research was supported by the Natural Sciences and Engineering Research Council of Canada. I am grateful to the coauthors of essays included in this collection: Scott Findlay (who made major contributions to three chapters), Abninder Litt, Daniel Saunders, Terry Stewart, and Jing Zhu. Please note that Daniel is first author of our joint article. Chris Eliasmith's exciting ideas about theoretical neuroscience contributed to several chapters. For comments or suggestions for particular chapters, I am indebted to him, William Bechtel, Chris Grisdale, Lloyd Elliott, Robert Hadley, Phil Johnson-Laird, Kostas Kampourakis, Eric Lormand, Elijah Millgram, Daniel Moerman, Nancy Nersessian, Eric Olsson, Robert Proctor, Peter Railton, David Rudge, Daniel Saunders, Cameron Shelley, and Terry Stewart. CBC Radio 2 provided the accompaniment.

I am grateful to my coauthors and to the respective publishers for permission to reprint the following essays:

Thagard, P., & Litt, A. (2008). Models of scientific explanation. In R. Sun (Ed.), *The Cambridge handbook of computational psychology* (pp. 549–564). Cambridge: Cambridge University Press. © Cambridge University Press.

Thagard, P. (2010). How brains make mental models. In L. Magnani, W. Carnielli & C. Pizzi (Eds.), *Model-based reasoning in science and technology: Abduction, logic, and computational discovery* (pp. 447–461). Berlin: Springer. © Springer.

Thagard, P., & Findlay, S. D. (2011). Changing minds about climate change: Belief revision, coherence, and emotion. In E. J. Olsson & S. Enqvist (Eds.), *Belief revision meets philosophy of science* (pp. 329–345). Berlin: Springer. © Springer.

Thagard, P. (2007). Coherence, truth, and the development of scientific knowledge. *Philosophy of Science, 74,* 28–47. © University of Chicago Press.

Thagard, P., & Stewart, T. C. (2011). The Aha! experience: Creativity through emergent binding in neural networks. *Cognitive Science, 35,* 1–33. © Cognitive Science Society.

Thagard, P. (forthcoming). Creative combination of representations: Scientific discovery and technological invention. In R. Proctor & E. J. Capaldi (Eds.), *Psychology of science.* Oxford: Oxford University Press. © Oxford University Press.

Saunders, D., & Thagard, P. (2005). Creativity in computer science. In J. C. Kaufman & J. Baer (Eds.), *Creativity across domains: Faces of the muse* (pp. 153–167). Mahwah, NJ: Lawrence Erlbaum. © Taylor and Francis.

Thagard, P. (2011). Patterns of medical discovery. In F. Gifford (Ed.), *Handbook of philosophy of medicine* (pp. 187–202). Amsterdam: Elsevier. © Elsevier.

Thagard, P. (2008). Conceptual change in the history of science: Life, mind, and disease. In S. Vosniadou (Ed.), *International handbook of research on conceptual change* (pp. 374–387). London: Routledge. © Taylor and Francis.

Thagard, P., & Findlay, S. (2010). Getting to Darwin: Obstacles to accepting evolution by natural selection. *Science & Education, 19,* 625–636. © Springer.

Thagard, P., & Zhu, J. (2003). Acupuncture, incommensurability, and conceptual change. In G. M. Sinatra & P. R. Pintrich (Eds.), *Intentional conceptual change* (pp. 79–102). Mahwah, NJ: Lawrence Erlbaum. © Taylor and Francis.

Thagard, P., & Findlay, S. (forthcoming). Conceptual change in medicine: Explanations of mental illness from demons to epigenetics. In W. J. Gonzalez (Ed.), *Conceptual revolutions: From cognitive science to medicine.* A Coruña, Spain: Netbiblo. © Netbiblo.

Finally, I am grateful to Judith Feldmann for skillful editing and to Eric Hochstein for help with the index.

I Introduction

1 What Is the Cognitive Science of Science?

Explaining Science

Science is one of the greatest achievements of human civilization, contributing both to the acquisition of knowledge and to people's well-being through technological advances in areas from medicine to electronics. Without science, we would lack understanding of planetary motion, chemical reactions, animal evolution, infectious disease, mental illness, social change, and countless other phenomena of great theoretical and practical importance. We would also lack many valuable applications of scientific knowledge, including antibiotics, airplanes, and computers. Hence it is appropriate that many disciplines such as philosophy, history, and sociology have attempted to make sense of how science works.

This book endeavors to understand scientific development from the perspective of cognitive science, the interdisciplinary investigation of mind and intelligence. Cognitive science encompasses at least six fields: psychology, neuroscience, linguistics, anthropology, philosophy, and artificial intelligence (for overviews, see Bermudez, 2010; Gardner, 1985; Thagard, 2005a). The main intellectual origins of cognitive science are in the 1950s, when thinkers such as Noam Chomsky, George Miller, Marvin Minsky, Allan Newell, and Herbert Simon began to develop new ideas about how human minds and computer programs might be capable of intelligent functions such as problem solving, language, and learning. The organizational origins of cognitive science are in the 1970s, with the establishment of the journal *Cognitive Science* and the Cognitive Science Society, and the first published uses of the term "cognitive science" (e.g., Bobrow & Collins, 1975).

Cognitive science has thrived because the problem of understanding how the mind works is far too complex to be approached using ideas and methods from only one discipline. Many researchers whose primary backgrounds are in psychology, philosophy, neuroscience, linguistics, anthropology, and computer science have realized the advantages of tracking work in some of the other fields of cognitive science. Many successful projects have fruitfully combined methodologies from multiple fields, for example, research on inference that is both philosophical and computational, research on language that is both linguistic and neuroscientific, and research on culture that is both anthropological and psychological.

Naturally, cognitive science has also been used to investigate the mental processes required for the practice of science. The prehistory of the cognitive science of science goes back to philosophical investigation of scientific inference by Francis Bacon, David Hume, William Whewell, John Stuart Mill, and Charles Peirce. Modern cognitive science of science began only in the 1980s when various psychologists, philosophers, and computer scientists realized the advantages of taking a multidisciplinary approach to understanding scientific thinking. Pioneers include: Lindley Darden, Ronald Giere, and Nancy Nersessian in philosophy; Bruce Buchanan, Pat Langley, and Herbert Simon in computer modeling; and William Brewer, Susan Carey, Kevin Dunbar, David Klahr, and Ryan Tweney in experimental psychology. Extensive references are given in the next section. The earliest occurrence of the phrase "cognitive science of science" that I have been able to find is in Giere (1987), although the idea of applying cognitive psychology and computer modeling to scientific thinking goes back at least to Simon (1966).

This chapter provides a brief overview of what the component fields of cognitive science bring to the study of science, along with a sketch of the merits of combining methods. It also considers alternative approaches to science studies that are often antagonistic to the cognitive science of science, including formal philosophy of science and postmodernist history and sociology of science. I will argue that philosophy, history, and sociology of science can all benefit from ideas drawn from the cognitive sciences. Finally, I give an overview of the rest of the book by sketching how the cognitive science of science can investigate some of the most important aspects of the development of science, especially explanation, discovery, and conceptual change.

Approaches to the Cognitive Science of Science

It would take an encyclopedia to review all the different approaches to science studies that have been pursued. Much more narrowly and concisely, this section reviews what researchers from various fields have sought to contribute to the cognitive science of science.

My own original field is the philosophy of science, and I described in the preface how concern with the structure and growth of scientific knowledge led me to adopt ideas and methods from psychology and artificial intelligence, generating books and articles that looked at different aspects of scientific thinking (e.g., Thagard, 1988, 1992, 1999, 2000). Independently, other philosophers have looked to cognition to enhance understanding of science, including Lindley Darden (1983, 1991, 2006), David Gooding (1990), Ronald Giere (1988, 1999, 2010), and Nancy Nersessian (1984, 1992, 2008). Andersen, Barker, and Cheng (2006), Magnani (2001, 2009), and Shelley (2003) also combine philosophy of science, history of science, and cognitive psychology. Collections of work on philosophical approaches to the cognitive science of science include Giere (1992) and Carruthers, Stich, and Siegal (2002).

Philosophy of science is not just a beneficiary of cognitive science but also a major contributor to it. Since the 1600s work of Francis Bacon (1960), philosophers have investigated the nature of scientific reasoning and contributed valuable insights on such topics as explanation (Whewell 1967), causal reasoning (Mill 1970), and analogy (Hesse 1966). Philosophy of science was sidetracked during the logical positivist era by (1) a focus on formal logic as the canonical way of representing scientific information and (2) a narrow empiricism incapable of comprehending the theoretical successes of science. Logical positivism was as inimical to understanding scientific knowledge as behaviorism was to understanding thinking in general.

In response to logical positivism, Russell Hanson (1958), Thomas Kuhn (1962), and others spurred interest among philosophers in the history of science, but there was a dearth of tools richer than formal logic for examining science, although Hanson and Kuhn occasionally drew on insights from Gestalt psychology. In the 1980s, when philosophers looked to cognitive science for help in understanding historical developments, we brought to the cognitive science of science familiarity with many aspects of high-level scientific thinking. The method that philosophy of science can most valuably contribute to the cognitive science of science consists in careful analysis of historical case studies.

Most psychologists concerned with scientific thinking adopt a very different method—behavioral experiments. Such experimentation is a crucial part of cognitive science, providing data about many different kinds of thinking that theories aim to explain. Professional scientists are rarely available for psychological experiments, but participants can be recruited from among the modern-day lab rats of cognitive psychologists—university undergraduates. Much of the valuable work on scientific thinking has been motivated by an attempt to understand how children can develop an understanding of science, a worthy enterprise that is part of both developmental and educational psychology.

Experimental and theoretical work on the development of scientific knowledge has been conducted by many psychologists (e.g., Carey, 1985, 2009; Dunbar, 1997, 2001; Dunbar & Fugelsang, 2005; Gentner et al., 1997; Klahr, 2000; Schunn & Anderson, 1999; Tweney, Doherty & Mynatt, 1981; Vosniadiou & Brewer, 1992). Like all cognitive scientists, psychologists can contribute to the development of theories about scientific thinking, but their main methodological contribution consists in behavioral experiments, although some psychologists such as Dedre Gentner and Ryan Tweney also undertake historical studies. Useful collections of work on the psychology of science include Crowley, Schunn, and Okada (2001), Gholson et al. (1989), Gorman et al. (2005), and Proctor and Capaldi (forthcoming). Other works in the psychology of science tied less closely to experimental cognitive psychology include Feist (2006), Simonton (1988), and Sulloway (1996). The introductory chapters below for parts II, III, and IV provide further references to work in the psychology of science on the more specific topics of explanation, discovery, and conceptual change.

In addition to philosophical/historical studies and behavioral experiments, the cognitive science of science has made extensive use of computational models, which have been theoretically and methodologically important since the 1950s. The theoretical usefulness comes from the fruitfulness of the hypothesis that thought is a kind of computation: thinking consists in applying processes to representations, just as computing consists in applying algorithms to data structures (see Thagard 2005a for a review). This hypothesis was far more powerful than previous attempts to understand the mind in terms of familiar mechanisms such as clockwork, vibrating strings, hydraulic systems, or telephone switchboards.

Moreover, computer modeling provides theorizing about the mind with a novel methodology—writing and running computer programs. Beginning with the seminal work on problem solving by Newell, Shaw, and Simon (1958), computer modeling has provided an invaluable tool for developing and testing ideas about mental processes (Sun, 2008b). Computational models of scientific thinking have been developed both by researchers in the branch of computer science called artificial intelligence, and by philosophers and psychologists who have adopted computer modeling as part of their methodological toolkit. There are many notable examples of computer simulations of different aspects of scientific thinking (e.g., Bridewell et al., 2008; Bridewell & Langley, 2010; Kulkarni & Simon, 1988, 1990; Langley et al., 1987; Lindsay et al., 1980; Shrager & Langley, 1990; Thagard, 1992; Valdes-Perez, 1995). The next section gives a more detailed discussion of how computational modeling contributes to the cognitive science of science.

Experimental neuroscience has so far made little contribution to understanding scientific thinking, even though it is becoming increasingly important to cognitive psychology and other areas such as social, developmental, and clinical psychology. The role of neuroscience in cognitive science has increased dramatically over the past two decades because of new technologies for observing neural activity using brain scanning tools such as functional magnetic resonance imaging (fMRI). The complementary theoretical side of neuroscience is the development of computational models that take seriously aspects of neural processing such as spiking neurons and interconnected brain areas that were neglected in the connectionist models of the 1980s (see, e.g., Dayan & Abbott, 2001; Eliasmith & Anderson, 2003). There has not yet emerged a distinct enterprise one would call the "neuroscience of science," although some of the kinds of thinking most relevant to scientific thought such as analogy, causal reasoning, and insight are beginning to receive experimental and theoretical investigation. What neuroscience can contribute to the understanding of science is knowledge about the neural processes that enable scientists to generate and evaluate ideas. Some of my own most recent work in chapters 3, 4, 8, and 19 employs ideas from theoretical neuroscience.

To complete this review of how the different fields of cognitive science contribute to the understanding of science, I need to include linguistics and anthropology. Unfortunately, I am not aware of much relevant research, although I can at least point to the work of Kertesz (2004) on the cognitive semantics of science, and to the work of Atran and Medin (2008) on folk concepts in biology across various cultures. Let me now return to why computer modeling is important for the cognitive science of science.

Methodology of Computational Modeling

What is the point of building computational models? One answer might come from the hypothetico-deductive view of scientific method, according to which science proceeds by generating hypotheses, deducing experimental predictions from them, and then performing experiments to see if the predicted observations occur. On this view, the main role of computational models is to facilitate deductions. There are undoubtedly fields such as mathematical physics and possibly economics where computer models play something like this hypothetico-deductive role, but their role in the cognitive sciences is much larger.

The hypothetico-deductive method is rarely applicable in biology, medicine, psychology, neuroscience, and the social sciences, where mathematically exact theories and precise predictions are rare. These sciences are better described by what I shall whimsically call the *mechanista* view of scientific method. Philosophers of science have described how many sciences aim for the discovery of mechanisms rather than laws, where a mechanism is a system of interacting parts that produce regular changes (e.g., Bechtel, 2008; Bechtel & Richardson, 1993; Bunge, 2003; Craver, 2007; Darden, 2006; Machamer, Darden, & Craver, 2000; Thagard, 2006a; Wimsatt, 2007). Biologists, for example, can rarely derive predictions from mathematically expressed theories, but they have been highly successful in describing mechanisms such as genetic variation and evolution by natural selection that have very broad explanatory scope. Similarly, I see cognitive science as primarily the search for mechanisms that can explain many kinds of mental phenomena such as perception, learning, problem solving, emotion, and language.

Computer modeling can be valuable for expressing, developing, and testing descriptions of mechanisms, at both psychological and neural levels of explanation. In contemporary cognitive science, theories at the psychological level postulate various kinds of mental representations and processes that operate on them to generate thinking. For example, rule-based theories of problem solving, from Newell and Simon (1972) to Anderson (2007), postulate (1) representations of goals and if-then rules and (2) search processes involving selection and firing of rules. The representations are the parts and the processes are the interactions that together provide a mechanism that explains mental changes that accomplish tasks. Other cognitive science theories can also be understood as descriptions of mechanisms, for example, connectionist models that postulate simple neuronlike parts and processes of spreading activation that produce mental changes (Rumelhart & McClelland, 1986). Computational neuroscience now deals with much more biologically realistic neural entities and processes than connectionism, but the aim is the same: to describe the mechanisms that explain neuropsychological phenomena.

Expressing and developing such theoretical mechanisms benefits enormously from computational models. It is crucial to distinguish between theories, models, and programs. On the mechanista view, a theory is a description of mechanisms, and a model is a simplified description of the mechanisms postulated to be responsible for some phenomena. In computational models, the simplifications consist of proposing general kinds of data structures and algorithms that correspond to the parts and interactions that the theory postulates. A computer program produces a still more specific and idealized account of the postulated parts and interactions using data structures and algorithms in a particular programming language. For example, the theory of problem solving as rule application using means-ends reasoning gets a simplified description in a computational model with rules and goals as data structures and means-ends search as interactions. A computer program implements the model and theory in a particular programming language such as LISP or JAVA that makes it possible to run simulations. Theoretical neuroscience uses mathematically sophisticated programming tools such as MATLAB to implement computational models of neural structures and processes that approximate to mechanisms that are hypothesized to operate in brains.

Rarely, however, do computer modelers proceed simply from theory to model to program in the way just suggested. Rather, thinking about how to write a computer program in a familiar programming language enables a cognitive scientist to express and develop ideas about what parts and interactions might be responsible for some psychological phenomena. Hence the development of cognitive theories, models, and programs is a highly interactive process in which theories stimulate the production of programs and vice versa. It is a mistake, however, to identify theories with programs, because any specific program will have many details arising from the peculiarities of the programming language used. Nevertheless, writing computer programs helps enormously to develop theoretical ideas expressed as computer models. The computer model provides a general analogue of the mechanisms postulated by the theory, and the program provides a specific, concrete, analogical instantiation of those mechanisms.

In the biological, social, and cognitive sciences, descriptions of mechanism are rarely so mathematical that predictions can be deduced, but running computer programs provides a looser way of evaluating theories and models. A computer program that instantiates a model that simplifies a theory can be run to produce simulations whose performance can be compared to actual behaviors, as shown in systematic observations, controlled behavioral experiments, or neurological experiments.

There are three degrees of evaluation that can be applied, answering the following questions about the phenomena to be explained:

1. Is the program capable of performing tasks like those that people have been observed doing?

2. Does the behavior of the program qualitatively fit with how people behave in experiments?

3. Does the behavior of the program quantitatively fit numerical data acquired in experiments?

Ideally, a computer program will satisfy all three of these tests, but often computer modeling is part of a theoretical enterprise that is well out in front of experimentation. In such cases, the program (and the model and theory it instantiates) can be used to suggest new experiments whose resulting data can be compared against the computer simulations. In turn, data that are hard to explain given currently available mechanisms may suggest new mechanisms that can be simulated by computer programs whose behaviors can once again be compared to those of natural systems. The three questions listed above apply to models of psychological

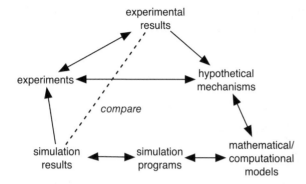

Figure 1.1
The role of computer models in developing and testing theories about mechanisms.
Lines with arrows indicate causal influences in scientific thinking. The dashed line
indicates the comparison between the results of experiments and the results of
simulations.

behavior, but analogous questions can be asked about computational simulations of neural data.

The general interactive process of mechanism-based theory development using computational models is shown in figure 1.1, which portrays
an interactive process with no particular starting point. Note that the
arrows between mechanisms and models, and between models and simulations, are symmetrical, indicating that models can suggest mechanisms
and programs can suggest models, as well as vice versa. In one typical
pattern, experimental results prompt the search for explanatory mechanisms that can be specified using mathematical–computational models
that are then implemented in computer programs. Simulations using these
programs generate results that can be compared with experimental results.
This comparison, along with insights gained during the whole process of
generating mechanisms, models, and simulations, can in turn lead to ideas
for new experiments that produce new experimental results.

Unified Cognitive Science Research

I have described philosophical, psychological, computational, and neuroscientific contributions to the understanding of science, but cognitive
science at its best combines insights from all of its fields. We can imagine

what an ideal research project in the cognitive science of science would be like, one beyond the scope of any single researcher except perhaps Herbert Simon. Consider a team of researchers operating with a core set of theoretical ideas and multiple methodologies. Let "ASPECT" stand for some aspect of scientific thinking that has been little investigated. We can imagine a joint enterprise in which philosophers analyze historical cases of ASPECT, psychologists perform behavioral experiments on how adults and children do ASPECT, neuroscientists perform brain scans of people doing ASPECT, and computational modelers write programs that can simulate ASPECT. Linguists and anthropologists might also get involved by studying whether ASPECT varies across cultures. Representatives of all six fields could work together to generate and test theories about the mental structures and processes that enable people to accomplish ASPECT. My own investigations into the cognitive science of science do not have anything like the scope of this imaginary investigation of ASPECT, but they variously combine different parts of the philosophical, historical, psychological, computational, and neuroscientific investigation of scientific thinking.

Unified investigations in the cognitive science of science can be normative as well as descriptive. It is sometimes said that philosophy is normative, concerned with how things ought to be, in contrast to the sciences which are descriptive, concerned with how things are. This division is far too simple, because there are many applied sciences, from engineering to medicine to clinical and educational psychology, that aim to improve the world, not just to describe it (Hardy-Vallée & Thagard, 2008). Conversely, if the norms that philosophy seeks to develop are to be at all relevant to actual human practices, they need to be tied to descriptions of how the world, including the mind, generally works. I have elsewhere defended the naturalistic view that philosophy is continuous with science, differing in having a greater degree of generality and normativity (Thagard, 2009, 2010a). This book assumes the priority of scientific evidence and reasoning over alternative ways of fixing belief such as religious faith and philosophical thought experiments, but I argue for that assumption in Thagard (2010a, ch. 2).

The cognitive science of science can take from its philosophical component and also from its applied components a concern to be normative as well as descriptive. An interdisciplinary approach to science can aim not only to describe how science works, but also to develop norms for how it might work better. The methodology is captured by the following norma-tive procedure (adapted from Thagard, 2010a, p. 211):

1. Identify a domain of practices, in this case ways of doing scientific research.

2. Identify candidate norms for these practices, such as searching for mechanisms.

3. Identify the appropriate goals of the practices in the given domain, such as truth, explanation, and technological applications.

4. Evaluate the extent to which different practices accomplish the relevant goals.

5. Adopt as domain norms those practices that best accomplish the rele-vant goals.

The descriptive side of cognitive science is essential for all of steps 1–4, but description can quickly lead to normative conclusions via the assessment shown in steps 4–5. My concern in the cognitive science of science is pri-marily descriptive, but normative issues will arise in chapters on climate change (ch. 5), truth (ch. 6), and values (ch. 17).

Other Approaches to Studying Science

Cognitive science is not the only way to study the practices and results of science, and there are alternative approaches that are antagonistic to it. Cognitive science is scorned by some philosophers, historians, and sociolo-gists who view it as fundamentally inadequate to understand the process of science and other important aspects of human life. I will now concisely review some of these alternatives, and describe why I think their opposi-tion misses the mark.

Within philosophy, the cognitive science of science exemplifies natural-ism, the view that philosophical deliberations should be tied to scientific evidence. Naturalistic philosophy has a venerable history, with practitio-ners such as Aristotle, Epicurus, Bacon, Locke, Hume, Mill, Peirce, Dewey, Quine, and many contemporary philosophers of science and mind. But philosophy also has a strong antinaturalistic strain, which challenges the relevance of science to philosophy from various directions. One prominent challenge seeks philosophical truths from reason alone, independent of scientific evidence; such truths are pursued by Plato, Kant, Frege, Husserl, and contemporary philosophers who try to use thought experiments to arrive at conceptual truths (for critiques of this approach, see Thagard, 2009, 2010a). This reason-based approach to philosophy tends to be antagonistic to cognitive science on the grounds that mind, like everything else, can be understood most deeply by methods that are a priori—independent of sense experience.

The antipsychologistic tendency of philosophy is also evident in contemporary work in the philosophy of science that employs formal methods such as symbolic logic, set theory, and probability theory. All of these tools are potentially relevant to the cognitive science of science, but formal philosophy of science uses them to the exclusion of many other tools (including the varied computational ones mentioned above) that cognitive science can bring to the examination of scientific knowledge. Formal philosophy of science follows in the tradition of the logical positivists in assuming that scientific theories are best viewed as abstract structures rather than as mental representations. Such abstractions are of limited use in understanding the actual practice of science and the details of the growth of scientific knowledge. It is particularly odd that the philosophy of science should ignore branches of science such as psychology that are highly relevant to understanding how science works, but the continuing influence of the Fregean, antipsychologistic strain of analytic philosophy is large.

A very different challenge to naturalistic philosophy comes from a more nihilistic direction that is generally skeptical of scientific and philosophical claims to achieve knowledge. Philosophers such as Nietzsche, Heidegger, Derrida, Foucault, and Lyotard rejected Enlightenment values of evidence, rationality, and objectivity. From a postmodernist perspective, science is just another human enterprise beset by power relations, whose discourse can be investigated by the same hermeneutic means that apply to other institutions. Cognitive science is then merely an attempt by scientists and science-oriented philosophers to exaggerate their own importance by privileging one style of thinking. In contrast, chapter 6 below provides a defense of scientific realism, the view that science aims to achieve truth and sometimes succeeds.

The postmodernist rejection of science as a way of knowing the world has infected much work in the history and sociology of science. Around the same time that the cognitive science of science was taking off, an alternative movement arose that managed to take over science studies programs at many universities. Sociologists of science produced a research program called the Sociology of Scientific Knowledge that abandoned the normative assessment of science in favor of purely sociological explanations of how science develops (e.g., Barnes, Bloor & Henry, 1996). Latour and Woolgar (1986) even called for a ten-year moratorium on cognitive explanations of science until sociologists had had a chance to explain all aspects of scientific development. That moratorium has long expired, and sociologists have obviously left lots of science to be explained. Moreover, some prominent proponents of postmodern sociology of science have made the shocking discovery that science and technology might even have *something* to do with reality (Latour, 2004).

In contrast to the imperialism of sociologists who think they can explain everything about scientific development, the cognitive science of science is friendly to sociological explanations. Power relations are undoubtedly an important part of scientific life, from the local level of laboratory politics to the national level of funding decisions. Like some analytic philosophers, some sociologists suffer from psychophobia, the fear of psychology, but cognitive approaches to science are compatible with the recognition of important social dimensions of science. For example, in my study of the development and acceptance of the bacterial theory of ulcers, I took into account social factors such as collaboration and consensus as well as psychological processes of discovery and evaluation (Thagard, 1999). Other works in the cognitive science of science have similarly attended to social dimensions (e.g., Dunbar, 1997; Giere, 1988). The cognitive and the social sciences should be seen as complements, not competitors, in a unified enterprise that might be called *cognitive social science*. Anthropology, sociology, politics, and economics can all be understood as requiring the integration of psychological and social mechanisms, as well as neural and molecular ones (Thagard, 2010d, forthcoming-c). Novel kinds of computer models are needed to explore how the behavior of groups can depend recursively on the behavior of individuals who think of themselves as members of groups. Agent-based models of social phenomena are being developed, but they are only just beginning to incorporate psychologically realistic agents (Sun, 2008a; Thagard, 2000, ch. 7, presents a cognitive-social model of scientific consensus). The aim of these models is not to reduce the social to the psychological and neural, but rather to show rich interconnections among multiple levels of explanation. My hope is that future work on cognitive-social interactions will provide ways of simulating social aspects of science using techniques under development (Thagard, forthcoming-c).

Studies in the Cognitive Science of Science

In the rest of this book, however, I largely neglect social factors in science in order to concentrate on its philosophical, psychological, computational, and neural aspects. Even within the cognitive realm, the investigations reported here are selective, dealing primarily with explanation, discovery, and conceptual change. I understand science broadly to include medicine and technology, which are discussed in several of the chapters.

Part II considers cognitive aspects of explanation and related scientific practices concerned with the nature of theories and theory choice. After a brief overview that makes connections to related work, four chapters develop cognitive perspectives on the nature of explanation, mental models, theory choice, and resistance to scientific change. Climate change provides a case study where normative models of theory acceptance based on explanatory coherence are ignored because of psychological factors. This part also includes the most philosophical chapter in the book, arguing that coherence in science sometimes leads to truth.

Part III concerns scientific discovery understood as a psychological and neural process. Formal philosophy of science and sociological approaches have had little to say about how discoveries are made. In contrast, this part contains a series of studies about the psychological and neural processes that lead to breakthroughs in science, medicine, and technology.

Part IV shows how discoveries of new theories and explanations lead to conceptual change, ranging from the mundane addition of new concepts to the dramatic reorganizations required by scientific revolutions. Four chapters describe conceptual change in the fields of biology, psychology, and medicine.

Finally, Part V presents two new essays concerned with the nature of values and with the neural underpinnings of scientific thinking. The chapter on values shows how the cognitive science of science can integrate descriptive questions about how science works with normative questions about how it ought to work. The final chapter builds on Chris Eliasmith's recent theory of semantic pointers to provide a novel account of the nature of scientific concepts such as *force, water,* and *cell.* (Please note that this book uses the following conventions: items in italics stand for concepts and items in quotes stand for words. For example, the concept *car* is expressed in English by the word "car" and applies to many things in the world, including my 2009 Honda Civic.)

The cognitive science of science inherits from the philosophy of science the problem of characterizing the structure and growth of scientific knowledge. It greatly expands the philosophical repertoire for describing the structure of knowledge by introducing much richer and empirically supported accounts of the nature of concepts, rules, mental models, and other kinds of representations. Even greater is the expansion of the repertoire of mechanisms for explaining the growth of scientific knowledge, through computationally rich and experimentally testable models of the nature of explanation, coherence, theory acceptance, inferential bias, concept formation, hypothesis discovery, and conceptual change. Adding an understanding of the psychological and neural processes that help to generate and establish scientific knowledge does not undercut philosophical concerns about normativity and truth, nor need it ignore the social processes that are also important for the development of scientific knowledge. Although the cognitive science of science is only a few decades old, I hope that the essays in this book, along with allied work by others, show its potential for explaining science.

II Explanation and Justification

2 Why Explanation Matters

Science attempts to answer many fundamental questions about how the world works. Why do objects fall when you drop them? Why does water become solid when it gets very cold? Why do animals have eyes? Why do people get sick? Answers to such questions are explanations that deepen our understanding by saying why things happen.

In ordinary life, explanations can be very shallow. I once saw a *Broom-Hilda* cartoon where the dialogue went something like this:

Irwin: Where does the Sun go at night?
Broom-Hilda: It's kept in a big barn in California.
Irwin: Then how does it get back East in the morning?
Broom-Hilda: By bus.
Irwin: Things are so easy to understand when you have a smart person to explain them!

This isn't much of an explanation, but it is a substantial question for philosophy and cognitive science to say what makes it inferior to, say, Newton's explanation that objects fall to the ground because of the force of gravity.

There are many reasons why explanation is an important topic for the cognitive science of science. First, explanation is clearly an important goal held by many scientists. Plato said that philosophy begins in wonder, and so does science. Many scientists are drawn into science by an intrinsic motivation to understand more about various aspects of physics, chemistry, biology, medicine, or the cognitive and social sciences. Such motivations are part of what keeps them in pursuit of new ideas. Scientific theories can be used for other purposes besides explanation, including prediction and practical applications, but these are complementary to explanation. Hence the cognitive science of science needs an explanation of explanation in order to understand a central aspect of scientific practice.

Second, explanation is an important part of science education. Teachers need to convey to students how science provides explanations and give them a taste of how scientific explanations work. Cognitive science should be able to integrate insights from philosophy, psychology, neuroscience, and computer modeling to elucidate what needs to go on in the minds of students as they learn at least to appreciate scientific explanations and at best to be able to develop new ones of their own.

Third, explanation is not only an intrinsic goal of science, but is intimately connected with another important goal—truth. Nowadays, people, especially social scientists, are sometimes embarrassed by the suggestion that science can achieve truth, interpreting talk of reality as a vestige of naive philosophical ideas that expired with Kant or with twentieth-century postmodernism. At the other extreme, some philosophers assume that science is primarily aimed just at truth, with explanation at best a sideshow. I reject both these views, and the chapters in Part II present a picture in which explanation is a key aspect of justifying the acceptance of hypotheses and theories. If theory choice is governed by inference to the best explanation, as chapter 5 assumes, then explanation is directly relevant to the question of what theories we should accept as true. Whether such theories really are true is a matter for discussion, as it is clear that many theories have been accepted by scientific communities that were later found to be defective. Chapter 6 provides a stronger connection between explanation, justification, and truth.

Of the fields of cognitive science, philosophy has the most ancient concern with the nature of explanation, going back at least to Aristotle. The philosophy of science took off in the 1800s with incisive discussions of explanation and explanatory reasoning by William Whewell, John Stuart Mill, and Charles Peirce. Peirce (1992) coined the term "abduction" for a kind of inference that generates and/or evaluates explanatory hypotheses. In the middle of the 1900s, the logical positivists developed a theory of explanation as deduction from laws that is still influential in philosophical circles (Hempel, 1965). More recently, many philosophers concerned with explanation in biology and cognitive science have highlighted the role that descriptions of mechanisms play in scientific explanations, as I reviewed in discussing the mechanista approach in chapter 1; see also Bechtel and Abrahamsen (2005). Woodward (2009) gives a good, brief overview of philosophical work on explanation. Kitcher and Salmon (1989) provide a useful older collection on the philosophy of scientific explanation.

Explanation has been an important concern for cognitive, developmental, and social psychologists. Cognitive psychologists have seen the relevance of explanation to the general theory of concepts, with some arguing that the functions of concepts include not just classification of objects but also explanation of why things happen (e.g., Lombrozo, 2009; Medin, 1989; Murphy, 2002). For example, saying that something is a bear can explain why it eats fish. Developmental psychologists have interpreted concept acquisition in children as partly aimed at providing explanations of why things happen (Gopnik, 1998; Keil, 2006). Social psychologists have long been concerned with how people explain the actions of others, a process they call attribution (Kelley, 1973). All of these investigations are relevant to scientific explanation, assuming some commonality between it and explanation in everyday life. See Keil and Wilson (2000) for a collection on the psychology of explanation.

Explanation was an important topic in artificial intelligence in the 1980s and 1990s (e.g., Minton et al., 1989), but it seems to have declined in importance as researchers moved to more statistical approaches. This decline is unfortunate, because AI needs to replicate the most sophisticated kinds of thinking, including what scientists do when they explain things. As chapter 3 describes, much work in AI has been limited to a deductive view of explanation, which at best captures only some kinds of scientific explanation.

Anthropologists and linguists have not, to my knowledge, done much to investigate the nature of scientific explanation. Nor has experimental and theoretical neuroscience yet said much about explanation, but some of the chapters below begin to fill this gap.

For Part II, I have chosen four recent papers as contributions to the cognitive science of explanation. Chapter 3 provides an overview of computational models of explanation, reviewing ones based on deduction, schemas, analogy, probability, and neural networks. It presents a model of how a simple form of abductive inference can be performed in a biologically plausible neural network. Chapter 4 shows how the same kind of neural network approach is relevant to explaining how high-level cognitive processes involving mental models can be understood neuro-computationally. This chapter also addresses the question of the extent to which cognition is embodied.

The next two chapters are concerned with philosophical questions about justification and truth, but they approach these from the cognitive perspective that explanation and inference are mental processes. Chapter 5 uses computational models of both correct and biased explanatory coherence to explain the nature of current debates about climate change. Ideally, claims about whether there is global warming and whether it is the result of human activities should be based solely on whether the hypotheses in question explain the evidence. These issues, however, are fraught with economic and political problems, so it is not surprising that people's thinking can be biased by their motivations. Cognitive science can explain not only how people think when they are doing it right, but also how thinking is often distorted by goals extraneous to the scientific aims of explanation and truth. Later, chapter 14 provides a similar account of resistance to Darwin's theory of evolution.

Finally, chapter 6, the most philosophical one in this book, argues that, under certain conditions, it is legitimate to conclude that theories that provide the best available evidence do indeed approximate the truth. This chapter provides a justification for the philosophical position called scientific realism, according to which science aims and sometimes succeeds in achieving truth. Thagard (2010a) provides a more general defense of realism.

3 Models of Scientific Explanation

Paul Thagard and Abninder Litt

Explanation

Explanation of why things happen is one of humans' most important cognitive operations. In everyday life, people are continually generating explanations of why other people behave the way they do, of why they get sick, of why computers or cars are not working properly, and of many other puzzling occurrences. More systematically, scientists develop theories to provide general explanations of physical phenomena such as why objects fall to Earth, chemical phenomena such as why elements combine, biological phenomena such as why species evolve, medical phenomena such as why organisms develop diseases, and psychological phenomena such as why people sometimes make mental errors.

This chapter reviews computational models of the cognitive processes that underlie these kinds of explanations of *why* events happen. It is not concerned with another sense of explanation that just means clarification, as when someone explains the U.S. Constitution. The focus will be on scientific explanations, but more mundane examples will occasionally be used, on the grounds that the cognitive processes for explaining why events happen are much the same in everyday life and in science, although scientific explanations tend to be more systematic and rigorous than everyday ones. In addition to providing a concise review of previous computational models of explanation, this chapter describes a new neural network model that shows how explanations can be performed by multimodal distributed representations.

Before proceeding with accounts of particular computational models of explanation, let us characterize more generally the three major processes involved in explanation and the four major theoretical approaches that have been taken in computational models of it. The three major processes are: providing an explanation from available information, generating new hypotheses that provide explanations, and evaluating competing explanations. The four major theoretical approaches are: deductive, using logic or rule-based systems; schematic, using explanation patterns or analogies; probabilistic, using Bayesian networks; and neural, using networks of artificial neurons. For each of these theoretical approaches, it is possible to characterize the different ways in which the provision, generation, and evaluation of explanations are understood computationally.

The processes of providing, generating, and evaluating explanations can be illustrated with a simple medical example. Suppose you arrive at your doctor's office with a high fever, headache, extreme fatigue, a bad cough, and major muscle aches. Your doctor will probably tell you that you have been infected by the influenza virus, with an explanation like:

People infected by the flu virus often have the symptoms you describe.
You have been exposed to and infected by the flu virus.
So, you have these symptoms.

If influenza is widespread in your community and your doctor has been seeing many patients with similar symptoms, it will not require much reasoning to provide this explanation by stating the flu virus as the likely cause of your symptoms.

Sometimes, however, a larger inferential leap is required to provide an explanation. If your symptoms also include a stiff neck and confusion, your doctor may make the less common and more serious diagnosis of meningitis. This diagnosis requires generating the hypothesis that you have been exposed to bacteria or viruses that have infected the lining surrounding the brain. In this case, the doctor is not simply applying knowledge already available to provide an explanation, but generating a hypothesis about you that makes it possible to provide an explanation. This hypothesis presupposes a history of medical research that led to the identification of meningitis as a disease caused by particular kinds of bacteria and viruses, research that required the generation of new general hypotheses that made explanation of particular cases of the disease possible.

In addition to providing and generating explanations, scientists and ordinary people sometimes need to evaluate competing explanations. If your symptoms are ambiguous, your doctor may be unsure whether you have influenza or meningitis, and therefore consider them as competing explanations of your symptoms. The doctor's task is then to figure out which hypothesis, that you have influenza or meningitis, is the *best* explanation of your disease. Similarly, at a more general level, scientific researchers had to consider alternative explanations of the causes of meningitis and select the best one. This selection presupposed the generation and provision of candidate explanations and involved the additional cognitive processes of comparing the candidates in order to decide which was most plausible.

Provision, generation, and evaluation of explanations can all be modeled computationally, but the forms these models take depend on background theories about what constitutes an explanation. One view, prominent in both philosophy of science and artificial intelligence, is that explanations are deductive arguments. An explanation consists of a deduction in which the explanatory target, to be explained, follows logically from the explaining set of propositions. Here is a simple example:

Anyone with influenza has fever, aches, and cough.
You have influenza.
So, you have fever, aches, and cough.

In this oversimplified case, it is plausible that the explanatory target follows deductively from the explaining propositions.

Often, however, the relation between explainers and explanatory targets is looser than logical deduction, and an explanation can be characterized as a causal schema rather than a deductive argument. A schema is a conceptual pattern that specifies a typical situation, as in the following example:

Explanatory pattern: Typically, influenza causes fever, aches, and cough.
Explanatory target: You have fever, aches, and cough.
Schema instantiation: Maybe you have influenza.

In medical research, the explanatory pattern is much more complex, as scientists can provide a much richer description of the genetic, biological, and immunological causes of infection. Like deductive explanations, schematic ones can be viewed as providing causes, but with a more flexible relation between explainers and what is explained.

Probability theory can also be used to provide a less rigid conception of explanation than logical deducibility. A target can be explained by specifying that it is probable given the state of affairs described by the explainers. In the flu case, the explanation has this kind of structure:

The probability of having fever, aches, and coughs given influenza is high. So influenza explains why you have fever, aches, and cough.

On this view, explanation is a matter of conditional probability rather than logical deducibility or schematic fit. Like the deduction and schema views, the probabilistic view of explanation has inspired interesting computational models, particularly ones involving Bayesian networks, which will be described below.

A fourth computational way of modeling explanation derives from artificial neural networks that attempt to approximate how brains use large groups of neurons, operating in parallel to accomplish complex cognitive tasks. The neural approach to explanation is not in itself a theory of explanation in the way that the deductive, schema, and probabilistic views are, but it offers new ways of thinking about the nature of the provision, generation, and evaluation of explanations. This quick overview sets the stage for the more detailed analysis of computational models of scientific explanation that follows. For a concise review of philosophical theories of explanation, see Woodward (2009); for more detail, see Kitcher and Salmon (1989).

Deductive Models

The view that explanations are deductive arguments has been prominent in the philosophy of science. According to Hempel (1965, p. 336), an explanation is an argument of the form:

$$C_1, C_2, \ldots C_k$$
$$\underline{L_1, L_2, \ldots, L_r}$$
$$E$$

Here the Cs are sentences describing particular facts, the Ls are general laws, and E is the sentence explained by virtue of being a logical consequence of the other sentences. This sort of explanation does occur in some areas of science such as physics, where laws stated as mathematical formulas enable deductive predictions.

Many computational models in artificial intelligence have presupposed that explanation is deductive, including ones found in logic programming, truth maintenance systems, explanation-based learning, qualitative reasoning, and in some approaches to abduction (a form of inference that involves the generation and evaluation of explanatory hypotheses). See, for example, Russell and Norvig (2003), Bylander et al. (1991), and Konolige (1992). These AI approaches are not intended as models of human cognition, but see Bringsjord (2008) for discussion of the use of formal logic in cognitive modeling.

Deductive explanation also operates in rule-based models, which have been proposed for many kinds of human thinking (Anderson, 1983, 1993, 2007; Holland et al., 1986; Newell & Simon, 1972; Newell, 1990). A rule-based system is a set of rules with an "if" part consisting of conditions (antecedents) and a "then" part consisting of actions (consequents). Rule-based systems have often been used to model human problem solving in which people need to figure out how to get from a starting state to a goal state by applying a series of rules. This is a kind of deduction, in that the application of rules in a series of if-then inferences amounts to a series of applications of the rule of deductive inference, modus ponens, which licenses inferences from p and *if p then q* to q. Most rule-based systems, however, do not always proceed from starting states to goal states, but can also work backward from a goal state to find a series of rules that can be used to get from the starting state to the goal state.

Explanation can be understood as a special kind of problem solving, in which the goal state is a target to be explained. Rule-based systems do not have the full logical complexity to express the laws required for Hempel's model of explanation, but they can perform a useful approximation. For instance, the medical example used in the introduction can be expressed by a rule like:

If X has influenza, then X has fever, cough, and aches.
Paul has influenza.
_____

Paul has fever, cough, and aches.

Modus ponens provides the connection between the rule and what is to be explained. In more complex cases, the connection would come from a sequence of applications of modus ponens as multiple rules get applied. In contrast to Hempel's account in which an explanation is a static argument, rule-based explanation is usually a dynamic process involving application of multiple rules. For a concrete example of a running program that accomplishes explanations in this way, see the PI ("processes of induction") model of Thagard (1988; code is available at http://cogsci .uwaterloo.ca). The main scientific example to which PI has been applied is the discovery of the wave theory of sound, which occurs in the context of an attempt to explain why sounds propagate and reflect.

Thus rule-based systems can model the provisions of explanations construed deductively, but what about the generation and evaluation of explanations? A simple form of abductive inference that generates hypotheses can be modeled as a kind of backward chaining. Forward chaining involves running rules forward in the deductive process that proceeds from the starting state toward a goal to be solved. Backward chaining occurs when a system works backward from a goal state to find rules that could produce it from the starting state. Human problem solving on tasks such as solving mathematics problems often involves a combination of forward and backward reasoning, in which a problem solver looks both at the how the problem is described and the answer that is required, attempting to make them meet. At the level of a single rule, backward chaining has the form: goal G is to be accomplished; there is the rule if A then G, that is, action A would accomplish G; so set A as a new subgoal to be accomplished. Analogously, people can backchain to find a possible explanation: fact F is to be explained; there is a rule if H then F, that is, hypothesis H would explain F; so hypothesize that H is true. Thus, if you know that Paul has fever, aches, and a cough, and you know the rule that if X has influenza, then X has fever, cough, and aches, then you can run the rule backward to produce the hypothesis that Paul has influenza.

The computational PI model performs this simple kind of hypothesis generation, but it also can generate other kinds of hypotheses (Thagard, 1988). For example, from the observation that the orbit of Uranus is perturbed, and the rule that if a planet has another planet near it then its orbit is perturbed, PI infers that there is some planet near Uranus; this is called "existential abduction." PI also performs abduction to rules that constitute the wave theory of sound: the attempt to explain why an arbitrary sound propagates generates not only the hypothesis that it consists of a wave but the general theory that all sounds are waves. PI also performs a kind of analogical abduction, a topic discussed in the next section on schemas.

Abductive inference that generates explanatory hypotheses is an inherently risky form of reasoning because of the possibility of alternative explanations. Inferring that Paul has influenza because it explains his fever, aches, and cough is risky because other diseases such as meningitis can cause the same symptoms. People should only accept an explanatory hypothesis if it is better than its competitors, a form of inference that philosophers call "inference to the best explanation" (Harman, 1973; Lipton, 2004). The PI cognitive model performs this kind of inference by taking into account three criteria for the best explanation: consilience, which is a measure of how much a hypothesis explains; simplicity, which is a measure of how few additional assumptions a hypothesis needs to carry out an explanation; and analogy, which favors hypotheses whose explanations are analogous to accepted ones. A more psychologically elegant way of performing inference to the best explanation, the ECHO model, is described below in the section on neural networks. Neither the PI nor the ECHO way of evaluating competing explanations requires that explanations be deductive.

In artificial intelligence, the term "abduction" is often used to describe inference to the best explanation as well as the generation of hypotheses. In actual systems, these two processes can be continuous, for example in the PEIRCE tool for abductive inference described by Josephson and Josephson (1994, p. 95). This is primarily an engineering tool rather than a cognitive model, but we mention it here as another approach to generating and evaluating scientific explanations, in particular medical ones involving interpretation of blood tests. The PEIRCE system accomplishes the goal of generating the best explanatory hypothesis by achieving three subgoals:

1. generation of a set of plausible hypotheses;

2. construction of a compound explanation for all the findings; and

3. criticism and improvement of the compound explanation.

PEIRCE employs computationally effective algorithms for each of these subgoals, but it does not attempt to do so in a way that corresponds to how people accomplish them.

Schema and Analogy Models

In ordinary life, and in many areas of science less mathematical than physics, the relation between what is explained and what does the explaining is usually looser than deduction. An alternative conception of this relation is provided by understanding an explanation as the application of a causal schema, which is a pattern that describes the relation between causes and effects. For example, cognitive science uses a general explanation schema with the following structure (Thagard, 2005a):

Explanation target: Why do people have a particular kind of **intelligent behavior**?

Explanatory pattern:
People have mental **representations**.
People have algorithmic **processes** that operate on those **representations**.
The **processes**, applied to the **representations**, produce the **behavior**.

This schema provides explanations when the terms shown in boldface are filled in with specifics, and subsumes schemas that describe particular kinds of mental representations such as concepts, rules, and neural networks. Philosophers of science have discussed the importance of explanation schemas or patterns (Kitcher, 1993; Thagard, 1999).

A computational cognitive model of explanation schemas was developed in the SWALE project (Schank, 1986; Leake, 1992). This project modeled people's attempts to explain the unexpected 1984 death of a racehorse, Swale. Given an occurrence, the program SWALE attempts to fit it into memory. If a problem arises indicating an anomaly, then the program attempts to find an explanation pattern stored in memory. The explanation patterns are derived from previous cases, such as other unexpected deaths. If SWALE finds more than one relevant explanation pattern, it evaluates them to determine which is most relevant to the intellectual goals of the person seeking understanding. If the best-explanation pattern does not quite fit the case to be explained, it can be tweaked (adapted) to provide a better fit, and the tweaked version is stored in memory for future use. The explanation patterns in SWALE's database included both general schemas such as *exertion + heart defect causes fatal heart attack* and particular examples, which are used for case-based reasoning, a kind of analogical thinking. Leake (1992) describes how competing explanation patterns can be evaluated according to various criteria, including a reasoner's pragmatic goals.

Explaining something by applying a general schema involves the same processes as explaining using analogies. In both cases, reasoning proceeds as follows:

Identify the case to be explained.

Search memory for a matching schema or case.

Adapt the found schema or case to provide an explanation of the case to be explained.

In deductive explanation, there is a tight logical relation between what is explained and the sentences that imply it, but in schematic or analogical explanation there need only be a roughly specified causal relation.

Falkenhainer (1990) describes a program, PHINEAS, that provides analogical explanations of scientific phenomena. The program uses Forbus's (1984) qualitative process theory to represent and reason about physical change, and is provided with knowledge about liquid flow. When presented with other phenomena to be explained such as osmosis and heat flow, it can generate new explanations analogically by computing similarities in relational structure, using the Structure Mapping Engine (Falkenhainer, Forbus & Gentner, 1989). PHINEAS operates in four stages: access, mapping/transfer, qualitative simulation, and revision. For example, it can generate an explanation of the behavior of a hot brick in cold water by analogy to what happens when liquid flows between two containers. Another computational model that generates analogical explanations is the PI system (Thagard, 1988), mentioned above, which simulates the discovery of the wave theory of sound by analogy to water waves.

Thus computational models of explanation that rely on matching schematic or analogical structures based on causal fit provide alternatives to models of deductive explanation. These two approaches are not competing theories of explanation, because explanation can take different forms in different areas of science. In areas such as physics that are rich in mathematically expressed knowledge, deductive explanations may be available. But in more qualitative areas of science and everyday life, explanations are usually less exact and may be better modeled by application of causal schemas or as a kind of analogical inference.

Probabilistic Models

Another, more quantitative way of establishing a looser relation than deduction between explainers and their targets is to use probability theory. Salmon (1970) proposed that the key to explanation is *statistical relevance*, where a property B in a population A is relevant to a property C if the probability of B given A and C is different from the probability of B given A alone: $P(B|A\&C) \neq P(B|A)$. Salmon later moved away from a statistical understanding of explanation toward a causal mechanism account (Salmon, 1984), but other philosophers and artificial intelligence researchers have focused on probabilistic accounts of causality and explanation. The core idea here is that people explain why something happened by citing the factors that made it more probable than it would have been otherwise.

The main computational method for modeling explanation probabilistically is Bayesian networks, developed by Pearl (1988, 2000) and other researchers in philosophy and computer science (e.g., Spirtes, Glymour & Scheines, 1993; Neapolitan, 1990; Glymour, 2001; Griffiths, Kemp & Tenenbaum, 2008). A Bayesian network is a directed graph in which the nodes are statistical variables, the edges between them represent conditional probabilities, and no cycles are allowed: you cannot have A influencing B and B influencing A. Causal structure and probability are connected by the Markov assumption, which says that a variable A in a causal graph is independent of all other variables that are not its effects, conditional on its direct causes in the graph (Glymour, 2003).

Bayesian networks are convenient ways for representing causal relationships, as in figure 3.1. Powerful algorithms have been developed for making probabilistic inferences in Bayesian networks and for learning causal relationships in these networks. Applications have included scientific examples, such as developing models in the social sciences (Spirtes, Glymour & Scheines, 1993). Bayesian networks provide an excellent tool for computational and normative philosophical applications, but the relevant question for this chapter is how they might contribute to cognitive modeling of scientific explanation.

The psychological plausibility of Bayesian networks has been advocated by Glymour (2001) and Gopnik et al. (2004). They show the potential for using Bayesian networks to explain a variety of kinds of reasoning and learning studied by cognitive and developmental psychologists. Gopnik

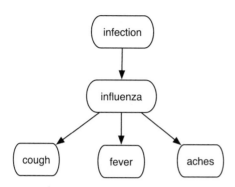

Figure 3.1
Causal map of a disease. In a Bayesian network, each node is a variable and the arrow indicates causality represented by conditional probability.

et al. (2004) argue that children's causal learning and inference may involve computations similar to those for learning Bayesian networks and for predicting with them. If they are right about children, it would be plausible that the causal inferences of scientists are also well modeled by Bayesian networks. From this perspective, explaining something consists in instantiating it in a causal network and using probabilistic inference to indicate how it depends causally on other factors. Generating an explanation consists in producing a Bayesian network, and evaluating competing explanations consists in calculating the comparative probability of different causes.

Despite their computational and philosophical power, there are reasons to doubt the psychological relevance of Bayesian networks. Although it is plausible that people's mental representations contain something like rough causal maps as depicted in figure 3.1, it is much less plausible that these maps have all the properties of Bayesian networks. First, there is abundant experimental evidence that reasoning with probabilities is not a natural part of people's inferential practice (Kahneman, Slovic & Tversky, 1982; Gilovich, Griffin & Kahneman, 2002). Computing with Bayesian networks requires a very large number of conditional probabilities that people not working in statistics have had no chance to acquire. Second, there is no reason to believe that people have the sort of information about independence that is required to satisfy the Markov condition and to make inference in Bayesian networks computationally tractable. Third, although it is natural to represent causal knowledge as directed graphs, in many scientific and everyday contexts such graphs should have cycles because of feedback loops. For example, marriage breakdown often occurs because of escalating negative affect, in which the negative emotions of one partner produce behaviors that increase the negative emotions of the other, which then produce behavior that increases the negative emotions of the first partner (Gottman et al., 2003). Such feedback loops are also common in biochemical pathways needed to explain disease (Thagard, 2003). Fourth, probability by itself is not adequate to capture people's understanding of causality, as argued in the last section of this chapter. Hence it is not at all obvious that Bayesian networks are the best way to model explanation by human scientists. Even in statistically rich fields such as the social sciences, scientists rely on an intuitive, nonprobabilistic sense of causality of the sort discussed below.

Neural Network Models

The most important approach to cognitive modeling not yet discussed here employs artificial neural networks. Applying this approach to high-level reasoning faces many challenges, particularly in representing the complex kinds of information contained in scientific hypotheses and causal relations. Thagard (1989) provided a neural network model of how competing scientific explanations can be evaluated, but did so using a localist network in which entire propositions were represented by single artificial neurons and in which relations between propositions are represented by excitatory and inhibitory links between the neurons. Although this model provides an extensive account of explanation evaluation, which is reviewed below, it reveals nothing about what an explanation is or how explanations are generated. Neural network modelers have been concerned mostly with applications to low-level psychological phenomena such as perception, categorization, and memory, rather than high-level ones such as problem solving and inference (O'Reilly & Munakata, 2000). However, this section shows how we can construct a neurologically complex model of explanation and abductive inference.

One benefit of attempting neural analyses of explanation is that it becomes possible to incorporate multimodal aspects of cognitive processing that tend to be ignored by the deductive, schematic, and probabilistic perspectives. Thagard (2007a) describes how both explainers and explanation targets are sometimes represented nonverbally. In medicine, for

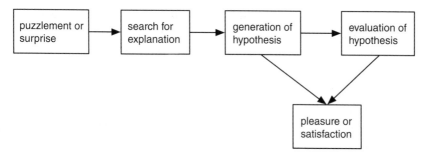

Figure 3.2
The process of abductive inference. (From Thagard, 2007a.)

example, doctors and researchers may employ visual hypotheses (say, about the shape and location of a tumor) to explain observations that can be represented using sight, touch, and smell as well as words. Moreover, the process of abductive inference has emotional inputs and outputs, because it is usually initiated when an observation is found to be surprising or puzzling, and it often results in a sense of pleasure or satisfaction when a satisfactory hypothesis is used to generate an explanation. Figure 3.2 provides an outline of this process. Let us now look at an implementation of a neural network model of this sketch.

The model of abduction described here follows the Neural Engineering Framework (NEF) outlined by Eliasmith and Anderson (2003), and is implemented using the MATLAB-based NEF simulation software *NESim*. The NEF proposes three basic principles of neural computation (Eliasmith & Anderson, 2003, p. 15):

1. Neural representations are defined by a combination of nonlinear encoding and linear decoding.

2. Transformations of neural representations are linearly decoded functions of variables that are represented by a neural population.

3. Neural dynamics are characterized by considering neural representations as control theoretic state variables.

These principles are applied to a particular neural system by identifying the interconnectivity of its subsystems, neuron response functions, neuron tuning curves, subsystem functional relations, and overall system behavior. For cognitive modeling, the NEF is useful because it provides a mathematically rigorous way of building more realistic neural models of cognitive functions.

The NEF characterizes neural populations and activities in terms of mathematical representations and transformations. The complexity of a representation is constrained by the *dimensionality* of the neural population that represents it. In rough terms, a single dimension in such a representation can correspond to one discrete "aspect" of that representation (e.g., speed and direction are the dimensional components of the vector quantity velocity). A hierarchy of representational complexity thus follows from neural activity defined in terms of one-dimensional scalars; vectors, with a finite but arbitrarily large number of dimensions; or functions, which are essentially *continuous* indexings of vector elements, thus ranging over infinite dimensional spaces.

The NEF provides for arbitrary computations to be performed in biologically realistic neural populations and has been successfully applied to phenomena as diverse as lamprey locomotion (Eliasmith & Anderson, 2003), path integration by rats (Conklin & Eliasmith, 2005), and the Wason card selection task (Eliasmith, 2005a). The Wason task model, in particular, is structured very similarly to the model of abductive inference discussed here. Both employ *holographic reduced representations* (HRRs), a high-dimensional form of distributed representation.

First developed by Plate (2003), HRRs combine the neurological plausibility of distributed representations with the ability to maintain complex, embedded structural relations in a computationally efficient manner. This ability is common in symbolic models and is often singled out as deficient in distributed connectionist frameworks; for a comprehensive review of HRRs in the context of the distributed versus symbolic representation debate, see Eliasmith and Thagard (2001). HRRs consist of high-dimensional vectors combined via multiplicative operations, and are similar to the tensor products used by Smolensky (1990) as the basis for a connectionist model of cognition. But HRRs have the important advantage of *fixed dimensionality*: the combination of two n-dimensional HRRs produces another n-dimensional HRR, rather than the $2n$ or even n^2 dimensionality one would obtain using tensor products. This avoids the explosive computational resource requirements of tensor products to represent arbitrary, complex structural relationships.

HRR representations are constructed through the multiplicative *circular convolution* (denoted by $\otimes$) and are decoded by the approximate inverse operation, *circular correlation* (denoted by #). The details of these operations are given in the appendixes of Eliasmith and Thagard (2001), but in general if $C = A \otimes B$ is encoded, then $C\#A \approx B$ and $C\#B \approx A$. The approximate nature of the unbinding process introduces a degree of noise, proportional to the complexity of the HRR encoding in question and in inverse proportion to the dimensionality of the HRR (Plate, 2003). As noise tolerance is a requirement of any neurologically plausible model, this loss of representation information is acceptable, and the "cleanup" method of recognizing encoded HRR vectors using the dot product can be used to find the vector that best fits what was decoded (Eliasmith & Thagard, 2001). Note that HRRs may also be combined by simple superposition (i.e., addition): $P = Q \otimes R + X \otimes Y$, where $P\#R \approx Q$, $P\#X \approx Y$, and so on. The operations required for convolution and correlation can be implemented in a recurrent connectionist network (Plate, 2003) and in particular under the NEF (Eliasmith, 2005a).

In brief, the new model of abductive inference involves several large, high-dimensional populations to represent the data stored via HRRs and learned HRR transformations (the main output of the model), and a smaller population representing emotional valence information (abduction only requires considering emotion scaling from surprise to satisfaction, and hence needs only a single dimension represented by as few as 100 neurons to represent emotional changes). The model is initialized with a base set of causal encodings consisting of 100-dimensional HRRs combined in the form

antecedent $\otimes$ *a* + *relation* $\otimes$ *causes* + *consequent* $\otimes$ *b*,

as well as HRRs that represent the successful explanation of a target *x* (*expl* $\otimes$ *x*). For the purposes of this model, only six different "filler" values were used, representing three such causal rules (*a* causes *b*, *c* causes *d*, and *e* causes *f*). The populations used have between 2,000 and 3,200 neurons each and are 100- or 200-dimensional, which is at the lower end of what is required for accurate HRR cleanup (Plate, 2003). More rules and filler values would require larger and higher-dimensional neural populations, an expansion that is unnecessary for a simple demonstration of abduction using biologically plausible neurons.

Following detection of a surprising *b*, which could be an event, proposition, or any sensory or cognitive data that can be represented via neurons, the change in emotional valence spurs activity in the output population toward generating a hypothesized explanation. This process involves employing several neural populations (representing the memorized rules and HRR convolution/correlation operations) to find an antecedent involved in a causal relationship that has *b* as the consequent. In terms of HRRs, this means producing (*rule # antecedent*) for ([*rule # relation* ≈ *causes*] and [*rule # consequent* ≈ *b*]). This production is accomplished in the 2,000-neuron, 100-dimensional output population by means of associative learning through recurrent connectivity and connection-weight updating (Eliasmith, 2005). As activity in this population settles, an HRR cleanup operation is performed to obtain the result of the learned transformation. Specifically, some answer is "chosen" if the cleanup result matches one encoded value significantly more than any of the others (i.e., is above some reasonable threshold value).

After the successful generation of an explanatory hypothesis, the emotional valence signal is reversed from surprise (which drove the search for an explanation) to what can be considered pleasure or satisfaction derived from having arrived at a plausible explanation. This in turn induces the output population to produce a representation corresponding to the successful dispatch of the explanandum *b*: namely, the HRR $expl_b = expl \otimes b$. Upon settling, it can thus be said that the model has accepted the hypothesized cause obtained in the previous stage as a valid explanation for the target *b*. Settling completes the abductive inference: emotional valence returns to a neutral level, which suspends learning in the output population and causes population firing to return to basal levels of activity.

Figure 3.3 shows the result of performing the process of abductive inference in the neural model, with activity in the output population changing with respect to changing emotional valence, and vice versa. The output population activity is displayed by dimension, rather than individual neuron, since the 100-dimensional HRR output of the neural ensemble as a whole is the real characterization of what is being represented. The boxed sets of numbers represent the results of HRR cleanups on the output population at different points in time; if one value reasonably dominates over the next few largest, it can be taken to be the "true" HRR represented by the population at that moment. In the first stage, the high emotional valence leads to the search for an antecedent of a causal rule" for *b*, the surprising explanandum. The result is an HRR cleanup best fitting to *a*, which is indeed the correct response. Reaching

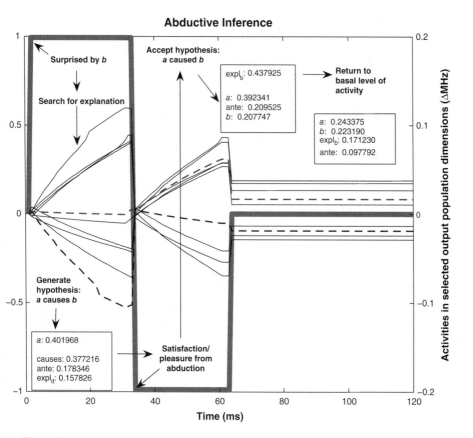

Figure 3.3

Neural activity in output population for abduction. For clarity, only a small (evenly spaced) selection of dimensional firing activities is displayed here (the full 2,000-neuron population has 100 dimensions). Activities for two specific population dimensions are highlighted by thickened dashed or dotted lines, to demonstrate the neural activity changes in response to changing emotional valence (shown as a thick solid line).

an answer with a reasonably high degree of certainty triggers an emotional valence shift (from surprise to satisfaction), which in turn causes the output population to represent the fact that b has been successfully explained, as represented by the HRR cleanup in the second stage of the graph. Finally, the emotional arousal shifts to a neutral state as abduction is completed, and the population returns to representing nothing particularly strongly in the final stage.

The basic process of abduction outlined previously (see fig. 3.2) maps very well to the results obtained from the model. The output population generates a valid hypothesis when surprised (since "a causes b" is the best memorized rule available to handle surprising b), and reversal of emotional valence corresponds to an acceptance of the hypothesis; hence the successful explanation of b.

In sum, the model of abduction outlined here demonstrates how emotion can influence the neural activity that underlies a cognitive process. Emotional valence acts as a *context gate* that determines whether the output neural ensemble must conduct a search for some explanation for surprising input, or whether some generated hypothesis needs to be evaluated as a suitable explanation for the surprising input.

The neural network model just described provides a mechanism for explanation, its emotional input and output, and a simple kind of abduction. It also performs a very simple sort of explanation evaluation, in that the causal rule that it selects from memory is chosen because it is a good match for the problem at hand, namely, explaining b. Obviously, however, this model is too simple to account for the comparative evaluation of explanatory theories as performed by the ECHO model (Thagard, 1989, 1992, 2000). In ECHO, hypotheses and pieces of evidence are represented by simple artificial neurons called units, which are connected by excitatory or inhibitory links that correspond to constraints between the propositions they represent. For example, if a hypothesis explains a piece of evidence, then there is a symmetric excitatory link between the unit that represents the hypothesis and the unit that represents the evidence. If two hypotheses contradict each other, then there is a symmetric inhibitory link between the two units that represent them. Units have activations that spread between them until the network reaches stable activation levels, which typically takes 60 to 100 iterations. If a unit ends up with positive activation, the proposition it represents is accepted, whereas if a unit ends up with negative activation, the proposition it represents is rejected.

ECHO has been used to model numerous cases in the history of science, and has also inspired experimental research in social and educational psychology (Read & Marcus-Newhall, 1993; Schank & Ranney, 1991). The model shows how a very high-level kind of cognition, evaluating complex theories, can be performed by a simple neural network performing parallel constraint satisfaction. ECHO has a degree of psychological plausibility, but for neurological plausibility it pales in comparison to the NEF model of abduction described earlier in this section. The largest ECHO model uses only around 200 units to encode the same number of propositions, whereas the NEF model uses thousands of spiking neurons to encode a few causal relations. Computationally, this seems inefficient, but of course the brain has many billions of neurons that provide its distributed representations.

How might one implement comparative theory evaluation as performed by ECHO within the NEF framework? Thagard and Aubie (2008) use the NEF to encode ECHO networks by generating a population of thousands of neurons. Parallel constraint satisfaction is performed by transformations of neurons that carry out approximately the same calculations that occur more directly in ECHO's localist neural networks. Hence it is now possible to model the evaluation of competing explanations using more biologically realistic neural networks.

Causality

Like most other models of explanation, these neural network models presuppose some understanding of causality. In one sense that is common in both science and everyday life, to explain something involves stating its cause. For example, when people have influenza, the virus that infects them is the cause of their symptoms such as fever. But what is a cause? Philosophical theories of explanation correlate with competing theories of causality; for example, the deductive view of explanation fits well with the Humean understanding of causality as constant conjunction. If all As are Bs, then someone can understand how being an A can cause and explain being a B. Unfortunately, universality is not a requisite of either explanation or causality. Smoking causes lung cancer, even though many smokers never get lung cancer, and some people with lung cancer have never smoked. Schematic models of explanation presuppose a primitive concept of causation without being able to say much about it. Probability theory may look like a promising approach to causality in that causes make their effects more probable than they would be otherwise, but such increased probability may be accidental or the result of some common cause. For example, the probability of someone drowning is greater on a day when much ice cream is consumed, but that is because of the common cause that more people go swimming and more people eat ice cream on hot days. Sorting out causal probabilistic information from misleading correlations requires much information about probability and independence that people usually lack.

Thagard (2007a) conjectured that it might be possible to give a neural network account of how organisms understand causality. Suppose, in keeping with research on infants' grasp of causality, that cause is a preverbal concept based on perception and motor control (Baillargeon, Kotovsky & Needham, 1995; Mandler, 2004). Consider an infant a few months old, lying on its back and swiping at a mobile suspended over its head. The infant has already acquired an image schema of the following form:

perception of situation + motor behavior ⇒ perception of new situation.

Perhaps this schema is innate, but alternatively it may have been acquired from very early perceptual–motor experiences in which the infant acted on the world and perceived the resultant changes. A simple instance of the schema would be:

stationary object + hand hitting object ⇒ moving object.

The idea of a preverbal image schema for causality is consistent with the views of some philosophers that manipulability and intervention are central features of causality (Woodward, 2004). The difference between A causing B and A merely being correlated with B is that manipulating A also manipulates B in the former case but not the latter. Conceptually, the concepts of manipulation and intervention seem to presuppose the concept of causation, because making something happen is on the surface no different from causing it to happen. However, although there is circularity at the verbal level, psychologically it is possible to break out of the circle by supposing that people have from infancy a neural encoding of the causality image schema described above. This nonverbal schema is the basis for understanding the difference between one event making another event happen and one event just occurring after the other.

The causality image schema is naturally implemented within the Neural Engineering Framework used to construct the model of abductive inference. Neural populations are capable of encoding both perceptions and motor behaviors, and are also capable of encoding relations between them. In the model of abductive inference described in the last section, *cause* (*c*, *e*) was represented by a neural population that encodes an HRR vector that captures the relation between a vector representing *c* and a vector representing *e*, where both of these can easily be nonverbal perceptions and actions as well as verbal representations. In the NEF model of abduction, there is no real understanding of causality, because the vector was generated automatically. In contrast, it is reasonable to conjecture that people have neural populations that encode the notion of causal connection as the result of their very early preverbal experience with manipulating objects. Because the connection is based on visual and kinesthetic experiences, it cannot be adequately formulated linguistically, but it provides the intellectual basis for the more verbal and mathematical characterizations of causality that develop later.

If this account of causality is correct, then a full cognitive model of explanation cannot be purely verbal or probabilistic. Many philosophers and cognitive scientists currently maintain that scientific explanation of phenomena consists in providing mechanisms that produce those phenomena (e.g., Bechtel & Abrahamsen, 2005; Sun, Coward & Zenzen, 2005). A mechanism is a system of objects whose interactions regularly produce changes. All of the computational models described in this chapter are mechanistic, although they differ in what they take to be the parts and interactions that are central to explaining human thinking; for the neural network approaches, the computational mechanisms are also biological ones. But an understanding of mechanism presupposes an understanding of causality, in that there must be a relation between the interactions of the parts that constitutes production of the relevant phenomenon. Because scientific explanation depends on the notion of causality, and because understanding of causality is in part visual and kinesthetic, future comprehensive cognitive models of explanation will need to incorporate neural network simulations of people's nonverbal understanding of causality.

Table 3.1
Summary of approaches to computational modeling of explanation.

	Target of explanation	Explainers	Relation between target and explainers	Mode of generation
Deductive	sentence	sentences	deduction	backward chaining
Schema	sentence	pattern of sentences	fit	search for fit, schema generation
Probabilistic	variable node	Bayesian network	conditional probability	Bayesian learning
Neural network	neural group: multimodal representation	neural groups	gated activation, connectivity	search, associative learning

Conclusion

This chapter has reviewed four major computational approaches to understanding scientific explanations: deductive, schematic, probabilistic, and neural network. Table 3.1 summarizes the different approaches to providing and generating explanations. To some extent, the approaches are complementary rather than competitive, because explanation can take different forms in different areas of science and everyday life. However, at the root of scientific and everyday explanation is an understanding of causality represented nonverbally in human brains by populations of neurons encoding how physical manipulations produce sensory changes. Another advantage of taking a neural network approach to explanation is that it becomes possible to model how abductive inference, the generation of explanatory hypotheses, is a process that is multimodal, involving not only verbal representations but also visual and emotional ones that constitute inputs and outputs to reasoning.

4 How Brains Make Mental Models

Introduction

Mental models are psychological representations that have the same relational structure as what they represent. They have been invoked to explain many important aspects of human reasoning, including deduction, induction, problem solving, language understanding, and human–machine interaction. But the nature of mental models and of the processes that operate on them has not always been clear from the psychological discussions. The main aim of this chapter is to provide a neural account of mental models by describing some of the brain mechanisms that produce them.

The neural representations required to understand mental models are also valuable for providing new understanding of how minds perform abduction, a kind of inference that generates and/or evaluates explanatory hypotheses. Considering the neural mechanisms that support abductive inference makes it possible to address several aspects of abduction, some first proposed by Charles Peirce, that have largely been neglected in subsequent research. These aspects include the generation of new ideas, the role of emotions such as surprise, the use of multimodal representations to produce "embodied abduction," and the nature of the causal relations that are required for explanations.

The suggestion that abductive inference is embodied raises issues that have been very controversial in recent discussions in psychology, philosophy, and artificial intelligence. This chapter argues that the role of emotions and multimodal representations in abduction supports a moderate thesis about the role of embodiment in human thinking, but not an extreme thesis that proposes embodied action as an alternative to the computational-representational understanding of mind.

Mental Models

How do you solve the following reasoning problem? Adam is taller than Bob, and Bob is taller than Dan; so what do you know about Adam and Dan? Readers proficient in formal logic may translate the given information into predicate calculus and use their encoding of the transitivity of "taller than" to infer that Adam is taller than Dan, via applications of the logical rules of universal instantiation, and-introduction, and modus ponens. Most people, however, report using a kind of image or model of the world in which they visualize Adam as taller than Bob and Bob as taller than Dan, from which they can simply "read off" the fact that Adam is taller than Dan.

The first modern statement of the hypothesis that minds use mechanical processes to model the world was by Kenneth Craik, who proposed that human thought provides a convenient small-scale model of a process such as designing a bridge (Craik, 1943 p. 59). The current popularity of the idea of mental models in cognitive science is largely due to Philip Johnson-Laird, who has used it extensively in explanations of deductive and other kinds of inference as well as many aspects of language understanding (e.g., Johnson-Laird, 1983, 2006; Johnson-Laird & Byrne, 1991). In his history of mental models, Johnson-Laird cites as an important precursor the ideas of Charles Peirce about the class of signs he called "likenesses" or "icons," which stand for things by virtue of a relation of similarity (Johnson-Laird, 2004). Earlier precursors may have been Locke and Hume with their view that ideas are copies of images. Many recent researchers have used mental models to explain aspects of thinking including problem solving (Gentner & Stevens, 1983), inductive learning (Holland et al., 1986), and human–machine interaction (e.g., Tauber & Ackerman, 1990). Hundreds of psychological articles have been published on mental models (see http://www.tcd.ie/Psychology/other/Ruth_Byrne/mental_models).

Nevertheless, our understanding of the nature of mental models has remained rather fuzzy. Nersessian (2008, p. 93) describes a mental model as a "structural, behavioral, or functional analog representation of a real-world or imaginary situation, event or process. It is analog in that it preserves constraints inherent in what is represented." But what is the exact nature of the psychological representations that can preserve constraints in the required way? One critic of mental model explanations of deduction dismisses them as "mental muddles" (Rips, 1986).

This chapter takes a new approach to developing the vague but fertile notion of mental models by characterizing them in terms of neural processes. A neural approach runs counter to the assumption of mental modelers such as Johnson-Laird and Craik that psychological explanation can proceed at an abstract functional and computational level, but I will try to display the advantages of operating at the neural as well as the psychological level. One advantage of a neural account of mental models is that it can shed new light on aspects of abductive inference.

Abduction

Magnani (1999) made the explicit connection between model-based reasoning and abduction, arguing that purely sentential accounts of the generation and evaluation of explanatory hypotheses are inadequate (see also Magnani, 2001; Shelley, 1996; Thagard & Shelley, 1997). A broader account of abduction, more in keeping with the expansive ideas of Peirce (1931–1958, 1992), can be achieved by considering how mental models that involve visual representations can contribute to explanatory reasoning. Sententially, abduction might be taken to be just "If p then q; why q? Maybe p." But much can be gained by allowing the p and q in the abductive schema to exceed the limitations of verbal information and include visual, olfactory, tactile, auditory, gustatory, and even kinesthetic representations. To take an extreme example, abduction can be prompted by a cry of "What's that awful smell?" that generates an explanation that combines verbal, visual, auditory, and motor representations into the answer that "Joe was trying to grate cheese onto the omelet but he slipped, cursed, and got some cheese onto the burner."

Moreover, there are aspects of Peirce's original descriptions of abduction that cannot be accommodated without taking a broader representational perspective. Peirce said that abduction is prompted by surprise, which is an emotion, but how can surprise be fitted into a sentential framework? Similarly, Peirce said that abduction introduces new ideas, but how could that happen in sentential schemas? Such ideas can generate flashes of insight, but both insight and their flashes seem indescribable in a sentential framework. Another problem concerns the nature of the "if-then" relation in the sentential abductive schema. Presumably it must be more than material implication, but what more is required? Logic-based approaches to abduction tend to assume that explanation is a matter of deduction, but philosophical discussions show that deduction is neither necessary nor sufficient for explanation (e.g., Salmon, 1989). I think that good explanations exploit causal mechanisms, but what constitutes the causal relation between what is explained and what gets explained? I aim to show that all of these difficult aspects of abduction—the role of surprise and insight, the generation of new ideas, and the nature of causality—can be illuminated by consideration of neural mechanisms.

Terminological note: Magnani (2009) writes of "non-explanatory abduction," which strikes me as self-contradictory. Perhaps there is a need for a new term describing a kind of generalization of abduction to cover other kinds of backward or inverse reasoning such as generating axioms from desired theorems, but let me propose to call this generalized abduction "gabduction" and retain the term "abduction" for Peirce's idea of the generation and evaluation of explanatory hypotheses.

Neural Representation and Processing

A full and rigorous description of current understanding of the nature of neural representation and processing is beyond the scope of this chapter, but I will provide an introductory sketch (for fuller accounts, see such sources as Churchland & Sejnowski, 1992; Dayan & Abbott, 2001; Eliasmith & Anderson, 2003; O'Reilly & Munakata, 2000; and Thagard, 2010a).

The human brain contains around 100,000,000,000 neurons, each of which has many thousands of connections with other neurons. These connections are either excitatory (the firing of one neuron increases the firing of the one it is connected to) or inhibitory (the firing of one neuron decreases the firing of the one it is connected to). A collection of neurons that are richly interconnected is called a neural population (or group, or ensemble). A neuron fires when it has accumulated sufficient voltage as the result of the firing of the neurons that have excitatory connections to it. Typical neurons fire around 100 times per second, making them vastly slower than current computers that operate at speeds of billions of times per second, but the massive parallel processing of the intricately connected brain enables it to perform feats of inference that are still far beyond the capabilities of computers.

A neural representation is not a static object like a word on paper or a street sign, but is rather a dynamic process involving ongoing change in many neurons and their interconnections. A population of neurons represents something by its pattern of firing. The brain is capable of a vast number of patterns: assuming that each neuron can fire 100 times per second, then the number of firing patterns of that duration is $(2^{100})^{100,000,000,000}$, a number far larger that the number of elementary particles in the universe, which is only about 10^{80}. I call this "Dickinson's theorem," after Emily Dickinson's beautiful poem "The Brain Is Wider Than the Sky." A pattern of activation in the brain constitutes a representation of something when there is a stable causal correlation between the firing of neurons in a population and the thing that is represented, such as an object or group of objects in the world (Eliasmith, 2005b; Parisien & Thagard, 2008). The claim that mental representations are patterns of firing in neural populations is a radical departure from everyday concepts and even from cognitive psychology until recently, but is increasingly supported by data acquired through experimental techniques such as brain scans and by rapidly developing theories about how brains work (e.g., Anderson, 2007; Smith & Kosslyn, 2007; Thagard 2010a).

Neural Mental Models

Demonstrating that neural representations can constitute mental models requires showing how they can have the same relational structure as what they represent, both statically and dynamically. Static mental models have spatial structure similar to what they represent, whereas dynamic mental models have similar temporal structure. Combined mental models capture both spatial and temporal structure, as when a person runs a mental movie that represents what happens in some complex visual situation such as two cars colliding.

The most straightforward kind of neural mental models are topographical sensory maps, for which Knudsen, du Lac, and Esterly (1987, p. 61) provide the following summary:

The nervous system performs computations to process information that is biologically important. Some of these computations occur in maps—arrays of neurons in which the tuning of neighboring neurons for a particular parameter value varies systematically. Computational maps transform the representation into a place-coded probability distribution that represents the computed values of parameters by sites of maximum relative activity. Numerous computational maps have been discovered, including visual maps of line orientation and direction of motion, auditory maps of amplitude spectrum and time interval, and motor maps of orienting movements.

The simplest example is the primary visual cortex, in which neighboring columns of neurons process information from neighboring small regions of visual space (Knudsen, du Lac & Esterly, 1987; Kaas, 1997). In this case, the spatial organization of the neurons corresponds systematically to the spatial organization of the world, in the same way that the location of major cities on a map of Brazil corresponds to the actual location of those cities.

Such topographic neural models are useful for basic perception, but they are not rich enough to support high-level kinds of reasoning such as the above "taller than" example. How populations of neurons can support such reasoning is still unknown, as brain scanning technologies do not have sufficient resolution to pin down neural activity in enough detail to inspire theoretical models of how high-level mental modeling can work. But let me try to extrapolate from current views on neural representation, particularly those of Eliasmith and Anderson (2003), to suggest how the brain might be able to make extra-topographic models of the world (see also Eliasmith, 2005b).

Neural populations can acquire the ability to encode features of the world as their firing activity becomes causally correlated with those features (A and B are causally correlated if they are statistically correlated as the result of causal interactions between A and B). Neural populations are also capable of encoding the activity of other neural populations, as the firing patterns of one population become causally correlated with the firing patterns of another population that feeds into it. If the input population is a topographic map, then the output population can become a more abstract representation of the features of the world, in two ways. The most basic retains some of the topographic structure of the input population, so that the output population is still a mental model of the world in that it shares some (but not all) relational structure with it. An even more abstract encoding is performed by an output neural population that captures key aspects of the encoding performed by the input population, but does so in a manner analogous to the way that language produces arbitrary, noniconic representations. Just as there is no similarity between the word "cat" and cats, so the output neural population may have lost the similarity with the original stimulus: not all thinking uses mental models. Nevertheless, in some cases the output population provides sufficient information to enable decoding that generates an inference fairly directly, as in the "taller-than" example. The encodings of Adam, Bob, and Dan that include their heights makes it possible to just "see" that Adam is taller than Dan.

A further level of representation is required for consciousness, such as the experienced awareness that Adam is taller than Dan. Many philosophers and scientists have suggested that consciousness requires representation of representation (for references, see Thagard & Aubie, 2008), but mental models seem to require several layers: representation of representation of representation. The conscious experience of an answer to a problem comes about because of activity in top-level neural populations that encode activity of medium-level modeling populations, that encode activity of low-level populations, that topographically represent features of the world. To put it another way, conscious models represent mid-level models that represent low-level topographic models that represent features of the world. The relation *representation* need not be transitive, but in this case it carries through, so that the conscious mental model represents the world and generates inferences about it.

So far, I have been focusing on mental models where the similar relation-structure is spatial, but temporal relations are just as important. When you imagine your national anthem sung by Michael Jackson, you are creating a mental model not only of the individual notes and tones but also of their sequence in time. Similarly, a mental model of a working device such as a windmill requires both visual-spatial representations of the blades and base of the windmill and also temporal representations of the blades of the mill. Not a lot of research has been done on how neurons can encode temporal relations, but I will explore two possibilities.

Elman (1990) and other researchers have shown how simple recurrent networks can encode temporal relationships needed for understanding language. A recurrent network is one in which output neurons feed back to provide input to the input neurons, producing a kind of temporal cycle that can retain information. Much more complex neural structures, however, would be needed to encode a song or running machine, perhaps something like the neural representation of a rule-based inference system being developed by Terry Stewart using Chris Eliasmith's neural engineering framework (Stewart & Eliasmith, 2009a). On this approach, a pattern of neural activation encodes a state of affairs that can be matched by a collection of rules capturing if-then relations. Running a temporal pattern is then a matter of firing off a sequence of rules, not by the usual verbal matching employed by rule-based cognitive architectures such as Anderson's (2007) ACT, but by purely neural network operations. If the neural populations representing the states of affairs are mental models of either the direct, topographic kinds or the abstracted, structure-preserving kinds, then the running of the rule-based system would constitute a temporal *and* spatial mental model of the world.

Generating New Ideas and Hypotheses

Peirce claimed that abduction could generate new ideas, but he did not specify how this could occur. If abduction is analyzed as a logical schema, then it is utterly mysterious how any new ideas could arise. The schema might be something like: "*q* is puzzling, *p* explains *q*, so maybe *p*." But this schema already includes the proposition *p*, so nothing new is generated. Hence logic-based approaches to abduction seem impotent to address what Peirce took to be a major feature of this kind of inference (cf. Thagard & Shelley, 1997). Thagard (1988) gave an account of how new concepts can be generated in the context of explanatory reasoning, but this account applies only to verbal concepts represented as frames with slots and values. In contrast, the view of representations as patterns of activity in neural populations can be used to describe the generation of new multimodal concepts. Chapter 8 shows how the mechanism of convolution in neural networks is capable of modeling not only the combination of representations but also the emotional reaction that successful combinations generate.

Creative conceptual combination does not occur randomly, but rather in the directed context of attempts to solve problems, including ones that require generation of new explanations. Chapter 3 described how abductive inference can operate with neural populations. At one level, our neural model of abduction is very simple, using thousands of neurons to model a transition from *q and p causes q* to *p*. The advantage in taking a neural approach to modeling, as I have already described, is that *p* and *q* need not be linguistic representations, but can operate in any modality. To take a novel example, q could be a neural encoding of pain that I feel in my finger, and p could be neural encoding of an image of a splinter in my finger. Then my abductive inference goes from the experience of pain to the adoption of the visual representation that there is a splinter.

Moreover, the neural model of abduction tracks the relevant emotions. Initially, the puzzling q is associated with motivating emotions such as surprise and irritation. But as the hypothesis p is abductively adopted, the emotional reaction changes to relief and pleasure. Thus neural modeling can capture emotional aspect of abductive reasoning.

But how can we understand the causal relation in "p causes q"? Chapter 3 (and Thagard, 2007a; Michotte, 1963) argued that causality should be construed not formally, as a deductive or probabilistic relation, but as a schema that derives from patterns of visual-motor experience. For example, when a baby discovers that moving its hand can move a rattle, it forms an association that combines an initial visual state with a subsequent motor and tactile state (pushing the rattle and feeling it) with a subsequent visual-auditory state (seeing and hearing the rattle move and make noise). Perhaps such sensory-motor-sensory schemas are innate, having been acquired by natural selection in the form of neural connections that everyone is born with; alternatively, they may be acquired very quickly by infants thanks to innate learning mechanisms. Moreover, the concept of force that figures centrally in many accounts of physical causality has its cognitive roots in body-based experiences of pushes and pulls (see ch. 18). Hence it seems appropriate to speak of "embodied abduction," since both the causal relation itself and the multimodal representations of many hypotheses and facts to be explained are tied to sensory operations of the human body. However, the topic of embodiment is highly controversial, so I now discuss how I think the embodiment of abduction and mental models needs to be construed.

Embodiment: Moderate and Extreme

I emphatically reject the extreme embodiment thesis that thinking is just embodied action and therefore incompatible with computational-representational approaches to how brains work (Dreyfus, 2007). I argue below that even motor control requires a high degree of representation and computation. Much more plausible is the moderate embodiment thesis that language and thought are inextricably shaped by embodied action, a view that is maintained by Gibbs (2006), Magnani (2009), and others. On this view, thinking still requires representations and computations, but the particular nature of these depends in part on the kind of bodies that people have, including their sensory and motor capabilities. My remarks about multimodal representations and the sensory-motor-sensory schemas that underlie causal reasoning provide support for the moderate embodiment thesis.

However, there are two main reasons for not endorsing the extreme embodiment thesis. First, many kinds of thinking, including causal reasoning, emotion, and scientific theorizing, take us well beyond sensorimotor processes, so explaining our cognitive capacities requires recognizing representational/computational abilities that outstrip embodied action. Second, even the central case of embodied action—motor control—requires substantial representational/computational capabilities.

I owe to Lloyd Elliott the following summary of why motor control is much harder than you might think. Merely reaching to pick up a book requires solutions to many difficult problems for the brain to direct an arm and hand to reach out and pick up the book. First, the signals that pass between the brain and its sensors and muscles are very noisy. Information about the size, shape, and location of the book is transmitted to the brain via the eyes, but the process of translating retinal signals into judgments about the book involved multiple stages of neural transformations (Smith & Kosslyn, 2007, ch. 2). Moreover, when the brain directs muscles to move the arm and hand in order to grasp the book, the signals sent involve noisy activity in millions of nerve cells.

Second, motor control is also made difficult by the fact that the context is constantly changing. You may need to pick up a book despite the fact that numerous changes are taking place, not only in the orientation of your body, but also in visual information such as light intensity and the presence of other objects in the area. A person can pick up a book even though another person has reached across to pick up another book. Third, there are unavoidable time delays as the brain plans and attempts to move the arm to pick up the book.

Fourth, motor control is not an automatic process that occurs instantly to people, but usually requires large amounts of learning. It takes years for babies to become adept at handling physical objects, and even adults require months or years to become proficient at difficult motor tasks such as playing sports. Fifth, motor control is not a simple linear process of the brain just telling a muscle what to do, but requires nonlinear integrations of the movements of multiple muscles and joints, which operate with many degrees of freedom. Picking up a book requires the coordination of all the muscles that move different parts of fingers, wrists, elbows, and shoulders.

Hence, grasping and moving objects is a highly complex task that has been found highly challenging by people attempting to build robots. Fortunately for humans, millions of years of animal evolution have provided us with the capacity to learn how to manipulate objects. Recent theoretical explanations of this capacity understand motor control as representational and computational, requiring mental models (see, e.g., Davidson & Wolpert, 2005; Wolpert & Ghahramani, 2000). What follows is a concise, simplified synthesis of their accounts.

The brain is able to manipulate objects because its learning mechanisms, both supervised and unsupervised, enable it to build powerful internal models of connections among sensors, brain, and world. A brain needs a *forward model* from movements to sensory results, which enables it to predict what will be perceived as the result of particular movements. It also needs an *inverse model* from sensory results to movements, which enables it to predict what movement will produce the desired perceived result. Forward and inverse models are both dynamic mental models in the sense I discussed earlier: the relational structure they share with what they represent is both spatial and temporal, concerning the location and movement of limbs to produce changes in the world. Motor control in general requires a high-level control process in which the brain enables the body to interact productively with the world through a combination of representations of situations and goals, forward and inverse models, perceptual filters, and muscle control processes. The overall process is highly complex and not at all like the kinds of manipulations of verbal symbols that some philosophers still take as the hallmark of representation and computation. But the brain's neural populations still stand for muscle movement and visual changes, with which their activity is causally correlated, so it is legitimate to describe the activities of such populations as representational. Moreover, the mental modeling, both forward and inverse, is carried out by systematic changes in the neural populations, and hence qualifies as "principled manipulation of representations" (Edelman, 2008, p. 29).

Let me summarize the argument in this section. Embodied action requires motor control. Motor control requires mental models, both forward and inverse, to identify dynamic relations among sensory information and muscle activity. Mental models are representational and computational. Hence embodied action requires representation and computation, so that it cannot provide an alternative to the representational/computational view of mind. Therefore, considerations of multimodal representations, embodied abduction, and sensorimotor conceptions of causality only support the moderate embodiment thesis, and in fact require a rejection of the extreme version.

Proponents of representation-free intelligence like to say that "the world is its own best model." As an advisory that a robot or other intelligent system should not need to represent everything to solve problems, this remark is useful; but taken literally it is clearly false. For imagining, planning, explaining, and many other important cognitive activities, the world is a very inadequate model of itself: far too complex and limited in its manipulability. In contrast, mental models operating at various degrees of abstraction are invaluable for high-level reasoning. The world might be its own best model if you're a cockroach, with very limited modeling abilities. But if you have the representational power of a human or powerful robot, then you can build simplified but immensely useful models of past and future events, as well as of events that your senses do not enable you to observe. Hence science uses abductive inference and conceptual combination to generate representations of theoretical (i.e., nonobservable) entities such as electrons, viruses, genes, and mental representations.

Cockroaches and many other animals are as embodied, embedded, and situated in the world as human beings, but they are far less effective than people at building science, technology, and other cultural developments. One of the many advantages that people gain from our much larger brains is the ability to work with mental models, including ones used for abduction and the generation of new ideas.

Conclusion

I finish with a reassessment of Peirce's ideas about abduction from the neural perspective that I have been developing. Peirce did most of his work on inference in the nineteenth century, well before the emergence of ideas about computation and neural processes. He was a scientist as well as a philosopher of science, and undoubtedly would have revised his views on the nature of inference in line with subsequent scientific developments.

On the positive side, first, Peirce was undoubtedly right about the importance of abduction as a kind of inference. The evaluative aspect of abduction is recognized in philosophy of science under the headings of inference to the best explanation and explanatory coherence, and the creative aspect is recognized in philosophy and artificial intelligence through work on how hypotheses are formed. Second, Peirce was prescient in noticing the emotional instigation of abduction as the result of surprise, although I do not know if he also noticed that achieving abduction generates the emotional response of relief. Third, Peirce was right in suggesting that the creation of new ideas often occurs in the context of abductive inference, even if abduction itself is not the generating process.

On the other hand, Peirce made several suggestions about abduction that do not fit well with current psychological and neural understanding of abduction. I do not think that emotion is well described as a kind of abduction, as it involves an extremely complex process that combines cognitive appraisal of a situation with respect to one's goals and perception of bodily states (Thagard & Aubie, 2008; Thagard, 2010a). At best, abductive inference is only a part of the broader parallel process of emotional reactions. Similarly, perception is not a kind of abduction, as it involves many more basic neuropsychological processes that are not well described as generation and evaluation of explanatory hypotheses (see, e.g., Smith & Kosslyn, 2007, ch. 2).

Finally, Peirce's suggestion that abduction requires a special instinct for guessing right is not well supported by current neuropsychological findings. Perhaps evolutionary psychologists would want to propose that there is an innate module for generating good hypotheses, but there is a dearth of evidence that would support this proposal. Rather, I prefer the suggestion of Quartz and Sejnowski (1997, 2002) that what the brain is adapted for is adaptability, through powerful learning mechanisms that humans can apply in many contexts. One of these learning mechanisms is abductive inference, which leads people to respond to surprising observations with a search for hypotheses that can explain them. Like all cognitive processes, this search must be constrained by contextual factors such as triggering conditions that cut down the number of new conceptual combinations that are performed (Thagard, 1988). Abduction and concept formation occur as part of the operations of a more general cognitive architecture.

I see no reason to claim that the constraints on these operations include preferences for particular kinds of hypotheses, which is how I interpret Peirce's instinct suggestion. Indeed, scientific abduction has led to the generation of many hypotheses that scientists now think are wrong (e.g., humoral causes of disease, phlogiston, caloric, and the luminiferous ether) and to many hypotheses that go against popular inclinations (e.g., Newton's force at a distance, Darwin's evolution by natural selection, Einstein's relativistic account of space-time, quantum mechanics, and the mind-brain identity theory). Although it is reasonable to suggest that the battery of innate human learning mechanisms includes ones for generating hypotheses to explain surprising events, there is no support for Peirce's contention that people must have an instinct for guessing *correctly*. Evolutionary psychologists like to compare the brain to a Swiss army knife that has many specific built-in capacities; but a more fertile comparison is the human hand, which evolved to be capable of many different operations, from grasping to signaling to thumb typing on smartphones. Peirce's view of abduction as requiring innate insight is thus as unsupported by current research as the view of Fodor (2000) that cognitive science cannot possibly explain abduction: many effective techniques have been developed by philosophers and AI researchers to explain complex causal reasoning.

I have tried to show in this chapter how Peirce's abduction is, from a neural perspective, highly consonant with psychological theories of mental models, which can also productively be construed as neural processes. Brains make mental models through complex patterns of neural firing and use them in many kinds of inference, from planning actions to the most creative kinds of abductive reasoning. I have endorsed a moderate thesis about the importance of embodiment for the kinds of representations that go into mental modeling, but critiqued the extreme view that sees embodiment as antithetical to mental models and other theories of representation. Further developments of neural theories of mental models should better clarify their roles in many important psychological phenomena.

5 Changing Minds about Climate Change: Belief Revision, Coherence, and Emotion

Paul Thagard and Scott Findlay

Scientific Belief Revision

Scientists sometimes change their minds. A 2008 survey on the Edge website presented more than 100 self-reports of thinkers changing their minds about scientific and methodological issues (http://www.edge.org/q2008/q08_index.html). For example, Stephen Schneider, a Stanford biologist and climatologist, reported how new evidence in the 1970s led him to abandon his previously published belief that human atmospheric emissions would likely have a cooling rather than a warming effect. Instead, he came to believe—what is now widely accepted—that greenhouse gases such as carbon dioxide are contributing to the dramatic trend of global warming. Similarly, Laurence Smith, a UCLA geographer, reported how in 2007 he came to believe that major changes resulting from global warming will come much sooner than he had previously thought. Observations such as the major sea-ice collapse in Canada's Northwest Passage had not been predicted to occur so soon by available computational models, but indicated that climate change is happening much faster than expected. Evidence accumulated over the past three decades is widely taken to show that global warming will have major impacts on human life, and that policy changes such as reducing the production of greenhouse gases are urgently needed. However, such scientific and policy conclusions have received considerable resistance, for example from former U.S. president George W. Bush and Canadian Prime Minister Stephen Harper.

A philosophical theory of belief revision should apply to the issue of global warming by explaining how most scientists have come to accept the following conclusions:

1. The Earth is warming.

2. Warming will have devastating impacts on human society.

3. Greenhouse gas emissions are the main causes of warming.

4. Reducing emissions is the best way to reduce the negative impacts of climate change.

In addition, the theory should provide insight not only into how scientists have come to adopt these beliefs, but also into why a few scientists and a larger number of leaders in business and politics have failed to adopt them.

We will show that belief revision about global warming can be modeled by a theory of explanatory coherence that has previously been applied to many cases of scientific belief change, including the major scientific revolutions (Thagard, 1992). We will present a computer simulation of how current evidence supports acceptance of important conclusions about global warming on the basis of explanatory coherence. In addition, we will explain resistance to these conclusions using a computational model of emotional coherence, which shows how political and economic goals can bias the evaluation of evidence and produce irrational rejection of claims about global warming.

Theory evaluation in philosophy, as in science, is comparative, and we will argue that explanatory coherence gives a better account of belief revision than major alternatives. The main competitors are Bayesian theories based on probability theory and logicist theories that use formal logic to characterize the expansion and contraction of belief sets. We will argue that the theory of explanatory coherence is superior to these approaches on a number of major dimensions. Coherence theory provides a detailed account of the rational adoption of claims about climate change, and can also explain irrational resistance to these claims. Moreover, we will show that it is superior to alternatives with respect to computational complexity. This chapter reviews the controversy about climate change, shows how explanatory coherence can model the acceptance of important hypotheses, and how emotional coherence can model resistance to belief revision. Finally, we will contrast the coherence account with Bayesian and logicist ones.

Climate Change

The modern history of beliefs about climate change began in 1896, when the Swedish scientist Svante Arrhenius discussed quantitatively the warming potential of carbon dioxide in the atmosphere (Arrhenius, 1896; Weart, 2003; see also http://www.aip.org/history/climate/index.html). The qualitative idea behind his calculations, now called the "greenhouse effect," had been proposed by Joseph Fourier in 1824: the atmosphere lets through the light rays of the Sun but retains the heat from the ground. Arrhenius calculated that if carbon dioxide emissions doubled from their 1896 levels, the planet could face warming of 5–6°C. But such warming was thought to be far off and even beneficial.

In the 1960s, after Charles Keeling found that carbon dioxide levels in the atmosphere were rising annually, Syukuro Manabe and Richard Wetherland calculated that doubling the carbon dioxide in the atmosphere would raise global temperatures a couple of degrees. By 1977, scientific opinion was coming to accept global warming as the primary climate risk for the next century. Unusual weather patterns and devastating droughts in the 1970s and 1980s set the stage for Congressional testimony by James Hansen, head of the NASA Goddard Institute for Space Studies. He warned that storms, floods, and fatal heat waves would result from the long-term warming trend that humans were causing. In 1988, the Intergovernmental Panel on Climate Change (IPCC) was established and began to produce a series of influential reports (http://www.ipcc.ch/). They concluded on the basis of substantial evidence that humans are causing a greenhouse effect warming, and that serious warming is likely to occur in the coming century. In 2006, former Congressman and presidential candidate Al Gore produced the influential documentary *An Inconvenient Truth*. This film, Gore's accompanying book, and the 2007 IPCC report helped solidify the view that major political action is needed to deal with the climate crisis (Gore, 2006).

Nevertheless, there remains substantial resistance to the IPCC's conclusions, in three major forms. Some scientists claim that observed warming can be explained by natural fluctuations in energy output by the Sun, the Earth's orbital pattern, and natural aerosols from volcanoes. Others are skeptical that human-emitted carbon dioxide can actually enhance the greenhouse effect. However, the most popular opposition today accepts most of the scientific claims but contends that there is no imminent crisis or necessity of costly actions.

Corporations and special interest groups fund research skeptical of a human-caused global warming crisis. Such works usually appear in nonscientific publications such as the *Wall Street Journal*, often with the financial backing of the petroleum or automotive industries and links to political conservatism. Corporations such as ExxonMobil spend millions of dollars funding right-wing think tanks and supporting skeptical scientists. For example, the Competitive Enterprise Institute (CEI) is a libertarian think tank that received $2.2 million between 1998 and 2006 from Exxon-Mobil. CEI sponsors the website globalwarming.org, which proclaims that policies being proposed to address global warming are not justified by current science and are a dangerous threat to freedom and prosperity.

The administration of U.S. President George W. Bush was highly influenced by the oil and energy industries consisting of corporations like ExxonMobil. The energy industry gave $48 million to Bush's 2000 campaign to become president, and has contributed $58 million in donations since then. Critics of the global warming crisis claim that there is a great deal of uncertainty associated with the findings of the IPCC: humans may not be the cause of recent warming. Moreover, such critics believe government should play a very small role in any emission regulation, as property rights and the free market will foster environmental responsibility. Their idea is that the power of technology will naturally provide solutions to global warming, and any emission cuts would be harmful to the economy of the United States, which is not obligated to lead any fight against global warming. Values that serve as the backbone to these beliefs include: small government, individual liberty and property rights, global equality, technology, and economic stability.

In contrast, global warming activists such as Al Gore believe that scientific predictions of global warming are sound and that the planet faces an undeniable crisis. Evidence shows that humans are the cause of warming, and the world's major governments must play a crucial leadership role in the changes necessary to save the planet. Some individuals have been convinced by a combination of evidence and moral motivations to switch views from the skeptics' to the environmentalists' camp. For example, the Australian global warming activist Tim Flannery was once skeptical of the case scientists had made for action against global warming. Influenced by evidence collected in the form of the ice cap record, he gradually revised his beliefs and became a prominent proponent of drastic actions to fight global warming (Flannery, 2006). Gore himself had to reject his childhood belief that the Earth is so vast and nature so powerful that nothing we do can have any major or lasting effect on the normal functioning of its natural systems. From the other direction, skeptics such as Bjørn Lomborg (2007) argue against the need for strong political actions to restrict carbon dioxide emissions. Let us now analyze the debate about global warming.

Coherence and Revision

The structure of the inferences that the Earth is warming because of production of greenhouse gases can be analyzed using the theory of explanatory coherence and the computer model ECHO. The theory and model have already been applied to a great many examples of inference in science, law, and everyday life (see, e.g., Thagard, 1989, 1992, 2000). The theory of explanatory coherence consists of the following principles:

Principle 1: Symmetry   Explanatory coherence is a symmetric relation, unlike, say, conditional probability. That is, two propositions p and q cohere with each other equally.

Principle 2: Explanation   (a) A hypothesis coheres with what it explains, which can be either evidence or another hypothesis; (b) hypotheses that together explain some other proposition cohere with each other; and (c) the more hypotheses it takes to explain something, the lower the degree of coherence.

Principle 3: Analogy   Similar hypotheses that explain similar pieces of evidence cohere.

Principle 4: Data priority   Propositions that describe the results of observations have a degree of acceptability on their own.

Principle 5: Contradiction   Contradictory propositions are incoherent with each other.

Principle 6: Competition   If p and q both explain a proposition, and if p and q are not explanatorily connected, then p and q are incoherent with each other (p and q are explanatorily connected if one explains the other or if together they explain something).

Principle 7: Acceptance   The acceptability of a proposition in a system of propositions depends on its coherence with them.

These principles do not fully specify how to determine coherence-based acceptance, but algorithms are available that can compute acceptance and rejection of propositions on the basis of coherence relations. The most psychologically natural algorithms use artificial neural networks that represent propositions by artificial neurons or *units* and represent coherence and incoherence relations by excitatory and inhibitory links between the units that represent the propositions. Acceptance or rejection of a proposition is represented by the degree of activation of the unit. The ECHO program spreads activation among all units in a network until some units are activated and others are inactivated, in a way that maximizes the coherence of all the propositions represented by the units. I will not present the technical details here, as they are available elsewhere (Thagard, 1992, 2000). Several different algorithms for computing coherence are analyzed by Thagard and Verbeurgt (1998).

The problem of scientific belief revision concerns how to deal with situations where new evidence or hypotheses generate the need to consider rejecting beliefs that have previously been accepted. According to the theory of explanatory coherence, belief revision should and often does proceed by evaluating all the relevant hypotheses with respect to all the evidence. A scientific database consists primarily of a set of propositions describing evidence and hypotheses that explain those propositions. There are coherence relations between pairs of propositions in accord with principle 1: when a hypothesis explains a piece of evidence, they cohere. There are also incoherence relations between pairs in accord with principles 5 and 6. When a new proposition comes along, representing either newly discovered evidence or a newly generated explanatory hypothesis, then this proposition is added to the overall set, along with positive and negative constraints based on the relations of coherence and incoherence that the new proposition has with the old ones. Then an assessment of coherence is performed in accord with principle 7, with the results telling you what to accept and what to reject. Belief revision takes place when a new proposition has sufficient coherence with the entire set of propositions that it becomes accepted and some proposition previously accepted becomes rejected.

Because a variety of algorithms is available for computing coherence, belief revision can be modeled in a highly effective and computationally efficient manner, involving substantial numbers of propositions. For example, Nowak and Thagard (1992a,b) simulated the acceptance of Copernicus's theory of the solar system and the rejection of Ptolemy's, with a total belief set of over 100 propositions. The LISP code for ECHO and various simulations is available at http://cogsci.uwaterloo.ca/Index.html. This site also makes available a partial JAVA version of ECHO.

Explanatory coherence is not intended to be a logically complete theory of belief revision, because it does not take into account a full range of operators such as conjunction and disjunction. Most emphatically, when a new proposition is added to a belief system, no attempt is made to add all its logical consequences, an infinitely large set beyond the power of any human or computer. Nevertheless, explanatory coherence gives a good account of what Gärdenfors (1988, 1992) describes as the three main kinds of belief change: expansion, revision, and contraction. Expansion takes place when a new proposition is introduced into a belief system, becoming accepted if and only if doing so maximizes coherence. Revision occurs when a new proposition is introduced into a belief system and leads other previously accepted propositions to be rejected because maximizing coherence requires accepting the new proposition and rejecting one or more old ones. Contraction occurs when some proposition becomes rejected because it no longer helps to maximize coherence. Simulations of these processes are described in the next section.

We do not use "maximize coherence" as a vague metaphor like most coherentist epistemologists, but rather as a computationally precise notion whose details are available elsewhere (e.g., Thagard, 1992, 2000). Logicist theories view belief revision as the result of expansion followed by contraction, but explanatory coherence computes expansion and contraction as happening at the same time in parallel.

Scientific belief revision comes in various degrees. A new proposition describing recently collected evidence may become accepted easily unless it does not fit well with accepted views. Such acceptance would be a simple case of expansion. However, if the new evidence is not easily explained by existing hypotheses, scientists may generate a new hypothesis to explain it. If the new hypothesis conflicts with existing hypotheses, either because it contradicts them or competes as an alternative hypothesis for other evidence, then major belief revision is required. Such revision may lead to theory change, in which one set of hypotheses is replaced by another set, as happens in scientific revolutions. The development of new ideas about climate change has not been revolutionary, since no major scientific theories have had to be rejected. But let us now look in more detail at how explanatory coherence can model the acceptance of global warming.

Simulating Belief Revision about Climate Change

To show how explanatory coherence can be used to simulate belief revision, we begin with a series of simple examples shown in figure 5.1. Simulation A shows the simplest possible case, where there is just one piece of evidence and one hypothesis that explains it. In accord with principle 4 of explanatory coherence, evidence propositions have a degree of acceptability on their own, which in the program ECHO is implemented by there being a positive constraint between each of them and a special unit EVIDENCE that is always accepted. Hence in simulation A, the acceptance of E1 leads to the acceptance of H1 as well.

Simulation B depicts a simple case of expansion beyond simulation A, in which a new piece of evidence is added and accepted. The situation gets more interesting in simulation C, where the hypothesis explains a predicted evidence proposition PE3, which, however, contradicts the actually observed evidence E4. A Popperian would say that H1 has now been

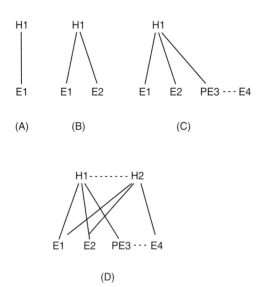

Figure 5.1
The straight lines indicate coherence relations (positive constraints) established because a hypothesis explains a piece of evidence. The dotted lines indicate incoherence relations (negative constraints). Coherence relations between an evidence unit and E1, E2, and E4 are not shown.

refuted and therefore should be rejected, but one failed prediction rarely leads to the rejection of a theory, for good reasons: perhaps the experiment that produced E4 was flawed, or there were other intervening factors that made the prediction of PE3 incorrect. Hence ECHO retains H1 while accepting E4 and rejecting PE3. Refutation of H1 requires the availability of an alternative hypothesis, as shown in simulation D. Here the addition of H2 provides an alternative explanation of the evidence, leading to its acceptance by virtue of its explanation of E4 as well as E1 and E2. This is a classic case of inference to the best explanation, where belief revision is accomplished through simultaneous expansion (the addition of H2) and retraction (the deletion of H1).

Belief revision about climate change can naturally be understood as a case of inference to the best explanation based on explanatory coherence. Figure 5.2 displays a drastically simplified simulation of the conflict between proponents of the view that climate change is caused by human activities and their critics. The hypothesis that is most important because it has major policy implications is that humans cause global warming. This hypothesis explains many observed phenomena, but figure 5.2 shows only two crucial generalizations from evidence: global temperatures are rising and the recent rise has been rapid. The main current alternative explanation is that rising temperatures are just the result of natural fluctuations in temperature that have frequently occurred throughout the Earth's

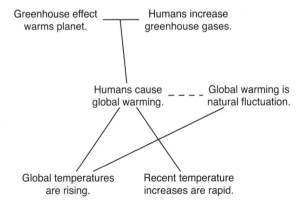

Figure 5.2
Highly simplified view of part of the controversy over climate change, with straight lines indicating coherence relations and dotted lines indicating incoherence ones.

history. Figure 5.2 also shows the favored explanation of how humans have caused global warming, through the greenhouse effect in which gases such as carbon dioxide and methane prevent energy radiation into space. Human industrial activity has produced huge increases in the amounts of such gases in the atmosphere over the past few hundred years.

Figure 5.2 shows only a small part of the explanatory structure of the controversy over climate change, and our full analysis is presented in figure 5.3, with more pieces of evidence and a fuller range of hypotheses. The input to our simulation using the program ECHO can be found in this chapter's appendix. The key competing hypotheses are GH3, that global warming is caused by humans, and NH4, that global warming is a natural fluctuation. As you would expect from the greater connectivity of hypothesis GH3 with the evidence, it wins out over NH4, which is rejected. The inputs to ECHO given in the appendix and the constraint structures shown in figure 5.3 capture much of the logical structure of the current debate

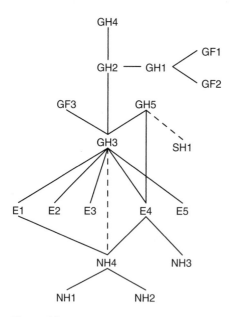

Figure 5.3
More detailed analysis of the controversy over climate change, with straight lines indicating coherence relations and dotted lines indicating incoherence ones. See the appendix to this chapter for a full description of the propositions and their relations.

over climate change. In accord with the current scientific consensus, ECHO accepts the basic claim that climate change is being caused by human activities that increase greenhouse gases.

ECHO can model belief revision in the previously skeptical by simulating what happens if only some of the evidence and explanations shown in figure 5.3 are available. For example, we have run a simulation that deletes most of the evidence for GH3 as well as the facts supporting GH1. In this case, ECHO finds NH1 more acceptable than GH3, rejecting the claim that humans are responsible for global warming. Adding back in the deleted evidence and the explanations of it by GH3 and the other global warming hypotheses produces a simulation of belief revision of the sort that would occur if a critic of human-caused warming were presented with more and more evidence. The result eventually is a kind of Gestalt switch, a tipping point in which the overall relations of explanatory coherence produce the adoption of new views and the rejection of old ones. Thus explanatory coherence can explain the move toward general acceptance of views currently dominant in the scientific community about climate change.

What explanatory coherence cannot explain is why some political and business leaders remain highly skeptical about global warming caused by human activities and the need to take drastic measures to curtail it. To understand their resistance, we need to expand the explanatory coherence model by taking into account emotional attitudes.

Simulating Resistance to Belief Revision

Scientific theory choice has the same logical structure as juror decisions in criminal cases. Just as scientists need to decide whether a proposed theory is the best explanation of the experimental evidence, juries need to decide whether the prosecutor's claim that the accused committed a crime is the best explanation of the criminal evidence. Ideally, juries are supposed to take into account all the available evidence and consider alternatives explanations of it. Often they do, but juries are like all people including scientists in having emotional biases that can lead to verdicts other than the one that provides the best explanation of the evidence. Thagard (2006, ch. 8) analyzed the decision of the jury in the O.J. Simpson case: explanatory coherence with respect to the available evidence should have led to the conclusion that Simpson was guilty, but the jury nevertheless acquitted him. However, the jurors' decision to acquit was simulated using the program HOTCO that simulates "hot coherence," which includes the contribution of emotional values to belief revision. Emotional values are a perfectly legitimate part of decision making as psychological indicators of the costs and benefits of expected outcomes. In the language of decision theory, deliberation requires utilities as well as probabilities. But normatively, belief revision should depend on evidence, not on utilities or emotional values.

We have already mentioned the motivations that lead some business and political leaders to be skeptical about claims about global warming. If climate change is a serious problem caused by human production of greenhouse gases, then measures need to be taken to curtail such production. Particularly affected by such measures will be oil companies, so it is not surprising that the research aimed at defusing alarm about global warming has been heavily supported by them. Moreover, some of the most powerful opposition to the Kyoto Protocol and other attempts to deal with global warming have come from politicians closely allied with the oil industry, such as U.S. president George W. Bush and Canadian Prime Minister Stephen Harper. In 2002, when he was leader of the Alberta-based Canadian Alliance, which later merged with the Conservative Party that he now leads, Harper wrote:

We're gearing up for the biggest struggle our party has faced since you entrusted me with the leadership. I'm talking about the "battle of Kyoto"—our campaign to block the job-killing, economy-destroying Kyoto Accord.

It would take more than one letter to explain what's wrong with Kyoto, but here are a few facts about this so-called "Accord":

—It's based on tentative and contradictory scientific evidence about climate trends.

—It focuses on carbon dioxide, which is essential to life, rather than upon pollutants.

—Canada is the only country in the world required to make significant cuts in emissions. Third World countries are exempt, the Europeans get credit for shutting down inefficient Soviet-era industries, and no country in the Western hemisphere except Canada is signing.

—Implementing Kyoto will cripple the oil and gas industry, which is essential to the economies of Newfoundland, Nova Scotia, Saskatchewan, Alberta and British Columbia.

—As the effects trickle through other industries, workers and consumers everywhere in Canada will lose. THERE ARE NO CANADIAN WINNERS UNDER THE KYOTO ACCORD.

—The only winners will be countries such as Russia, India, and China, from which Canada will have to buy "emissions credits." Kyoto is essentially a socialist scheme to suck money out of wealth-producing nations. (http://www.thestar.com/article/ 176382)

Prime Minister Harper has since moderated his position on global warming, as has George W. Bush, but both have been slow to implement any practical changes.

We conjecture that at the root of opposition to claims about global warming are the following concerns. Dealing with climate change would require government intervention to restrict oil usage, which is doubly bad for a conservative politician with a preference for free market solutions and a long history of association with oil-producing companies. Figure 5.4 expands figure 5.2 to include the strong emotional values of avoiding limiting oil use and production and avoiding government intervention in the economy. When the explanatory coherence relations shown in the

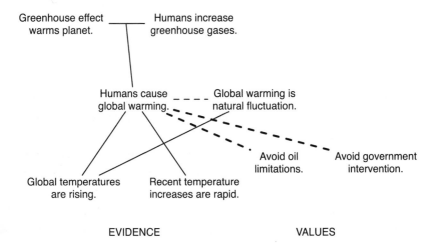

Figure 5.4
View of the controversy over climate change including emotional constraints as well as explanatory ones. As in previous figures, the solid lines indicate positive constraints based on explanatory relations and the thin dotted line indicates a negative constraint based on incompatibility. The thick dotted lines indicate negative emotional constraints.

appendix and figure 5.3 are supplanted with these values, belief revision slows down, so that more evidence is required to shift from rejection to acceptance of the hypothesis that global warming is being caused by human activities.

In the ECHO simulation of just the hypotheses and evidence shown in figure 5.4, the obvious result is the acceptance of the hypothesis that global warming is caused by humans, implying that political action can and should be taken against it. But we have also used the extended program HOTCO to yield a different result. If, as figure 5.4 suggests, the hypothesis that humans caused global warming is taken to conflict with the values of avoiding oil limitations and government intervention, then the simulation results in the rejection of human causes for global warming and acceptance of the alternative hypothesis of natural fluctuation. Of, course, the actual psychological process of motivated inference in this case is much more complex than figure 5.4 portrays, leading skeptics to challenge evidence and explanations as well as hypotheses. But figure 5.4 and the HOTCO simulation of it show how values can interfere with belief revision by undermining hypotheses that are better supported by the evidence. Hence a psychologically natural extension of ECHO can explain resistance to belief revision.

Alternative Theories of Belief Revision

Explanatory coherence is not the only available account of scientific belief revision. We also have at least the following: Popper's conjectures and refutations, Hempel's confirmation theory, Kuhn's paradigm shifts, Lakatos's methodology of research programs, and social constructionist claims that scientists revise their beliefs only to increase their own power. These are all much vaguer than the theory of explanatory coherence and its computational implementation in ECHO. Among formally exact characterization of belief revision, the two most powerful approaches are Bayesian ones that explain belief change using probability theory, and logicist ones that use ideas about logical consequence in deductive systems.

Detailed comparisons between explanatory coherence and Bayesian accounts of belief revision have been presented elsewhere (Thagard, 2000, ch. 8; Eliasmith & Thagard, 1997; Thagard, 2004; ch. 3, above). It should be feasible to produce a Bayesian computer simulation of the full climate change case shown in figure 5.3 and the appendix. A simulator such as JavaBayes (http://www.cs.cmu.edu/~javabayes/Home/) could be used to produce a Bayesian alternative to our ECHO simulator, as was done for a complex legal example by Thagard (2004). But doing so would require a very large number of conditional probabilities whose interpretation and provenance are highly problematic. In the simplest case where you have two hypotheses, H1 and H2, explaining a piece of evidence E1, JavaBayes would require specification of eight conditional probabilities, such as P(E1/H1 & not-H2). For example, the simplified model shown in figure 5.2 would require specification of P(global temperatures are rising/humans cause global warming & global warming is not a natural fluctuation) as well as seven other conditional probabilities. In general, the nodes in an explanatory coherence network can be translated into nodes in a Bayesian network with the links translated into directional arrows. The price to be paid is that for each node that has n arrows coming into it, it is necessary to specify 2^{n+1} conditional probabilities. In our ECHO simulation specified in the appendix, a highly connected node such as E4, which has three global warming hypotheses and four alternative hypotheses explaining it, would require specification of $2^8 = 256$ conditional probabilities, none of which can be estimated from available data. To produce a JavaBayes alternative to our full ECHO simulation, thousands of conditional probabilities would simply have to be made up. Bayesian models are very useful for cases where there are large amounts of statistical data, such as sensory processing in humans and robots; but they add nothing to understanding of cases of belief revision such as climate change, where probabilities are unknown.

Writers such as Olsson (2005) have criticized coherence theories for not being compatible with probability theory, but probability theory just seems irrelevant in qualitative cases of theory change. The comprehensive report of the Intergovernmental Panel on Climate Change sensibly relies on qualitative characterizations such as "extremely likely" and "very high confidence." Probability theory should be used whenever appropriate for statistics-based inference, but applying it to qualitative cases of causal reasoning such as climate changes obscures more than it illuminates.

The other major formal approach to belief revision uses techniques of formal logic to characterize the expansion and contraction of belief sets (see, e.g., Gärdenfors, 1988, 1992; Tennant, 1994, 2006). We will not attempt a full discussion here, but the explanatory coherence approach seems to us to be superior to logicist approaches in several respects, which we will briefly describe.

First, the explanatory coherence approach is both broad and deep, having been applied in detail to many important cases of scientific belief revision. In this chapter, we have shown its applicability to a belief revision problem of great contemporary interest, climate change. To our knowledge, logicist approaches to belief revision have not served to model any historical or contemporary cases of belief change. Explanatory coherence has less generality than logicist approaches, because it does not attempt to provide algorithms for computing revision and contraction functions for an *arbitrary* belief system. Rather, it focuses on revision in systems of hypotheses and evidence, which suffices for the most important kinds of belief change in science, criminal cases, and everyday life.

Second, the explanatory coherence approach has philosophical advantages over logicist approaches. It does not assume that changes in belief systems should be *minimal*, retaining as much as possible from our old beliefs. The principle of minimal change has often been advocated (e.g., by Gärdenfors, Tennant, Harman, and Quine) but rarely defended, and Rott (2000) has argued that it is not even appropriate for logicist approaches. Aiming for minimal change in belief systems seems to us no more justifiable than aiming for minimal change in political and social systems. Just as political conservatism should not serve to block needed social changes, so epistemic minimalism should not get in the way of needed belief changes. Explanatory coherence theory shows how to make just the changes that are needed to maximize the coherence of evidence and hypotheses. As long as they are productive, both epistemic and social innovations are to be valued. Just as aiming for minimal change in production of greenhouse gases may prevent dealing adequately with forthcoming climate crises, so aiming for minimal change in belief revision may prevent arriving at maximally coherent and arguably true theories. Coherence approaches have often been chided for neglecting the importance of truth as an epistemic goal, but chapter 6 argues that applications of explanatory coherence in science do in fact lead to truth when they produce the adoption of theories that are both broad (explaining much evidence) and deep (possessing underlying mechanistic explanations).

Third, explanatory coherence has computational advantages over logicist approaches that assume that a belief set is logically closed. Then belief sets are infinite, and cannot be computed by any finite machine. The restricted problem of belief contraction in finite belief sets has been shown to be NP-complete (Tennant, 2003). Problems in NP are usually taken by computer scientists to present computational difficulties, in that polynomial-time solutions are not available so they are characterized as intractable. It might seem that the coherence approach is in the same awkward boat, as Thagard and Verbeurgt (1998) showed that the coherence problem is NP-hard. However, many hard computational problems become tractable if a problem parameter is fixed or bounded by a fixed value (Gottlob, Scarcello & Sideri 2002; van Rooij, 2008). Van Rooij has shown that coherence problems become fixed-parameter tractable if they have a high ratio of positive to negative constraints. Fortunately, all programmed examples of explanatory coherence have a high positive-negative ratio, as you would expect from the fact that most constraints are generated by explanatory coherence principle 2. In particular, our simulation of the climate change case using the input in the appendix generates 10 negative constraints and 53 positive ones. Hence explanatory coherence in realistic applications appears to be computationally tractable.

Fourth, explanatory coherence has major psychological advantages over logistic approaches, which are intended to model not how people actually perform belief revision, but rather how belief revision should ideally take place. Explanatory coherence theory has affinities with a whole class of coherence models that have been applied to a wide range of psychological phenomena (Thagard, 2000). The psychological plausibility of explanatory coherence over logicist approaches will not be appreciated by those who like their epistemology to take a Platonic or Fregean form, but it is a clear advantage for those of us who prefer a naturalistic approach to epistemology. A related psychological advantage is that explanatory coherence meshes with psychological accounts that can explain what people are doing when they irrationally resist belief revision, as we saw in the case of political opposition to theories of climate change.

Conclusion

Our brief discussions of Bayesian and logicist accounts of belief revision hardly constitute a refutation of these powerful approaches, but they should suffice to indicate how they differ from the explanatory coherence approach. We have shown how explanatory and emotional coherence can illuminate current debates about climate change. We can use it to understand the rational adoption of the hypothesis that global warming is caused by human activities. Moreover, deviations from rational belief revision in the form of emotion-induced rejection of the best explanation of a wide range of evidence can be understood in terms of intrusion of emotional political values into the assessment of the best explanation.

Scientific belief revision is not supposed to be impeded by emotional prejudices, but such impedance is common. The acceptance of Darwin's theory of evolution by natural selection was slowed in the nineteenth century by fears of its threat to established theological theories of the creation of species. Such resistance continues today, as many people—but very few scientists—continue to reject evolutionary theory as incompatible with their religious beliefs and desires. In practical settings, encouraging belief change about Darwin's theories, as about climate change, requires dealing with emotional constraints as well as cognitive ones generated by the available evidence, hypotheses, and explanations. Thus a broadly applicable, computationally feasible, and psychologically insightful account of belief revision such as that provided by the theory of explanatory coherence should be practically useful as well as philosophically informative.

Appendix

Input to the ECHO Simulation of the Acceptance of the Claim That Global Warming Is Caused by Humans

Global warming: A simplified model of anthropogenic forcing vs. natural causes
Evidence

E1. Average global temperatures have risen significantly since 1880.

E2. The rate of warming is rapidly increasing.

E3. The recent warming is more extreme than any other warming period as far back as the record shows to 1000 AD.

E4. Arctic ice is rapidly melting and glaciers around the world are retreating.

E5. Global temperature shows strong correlation with carbon dioxide levels throughout history.

IPCC/Gore's facts

GF1. Carbon dioxide, methane gas, and water vapor are greenhouse gases.

GF2. Greenhouse gases absorb infrared radiation, some of which is reemitted back to the Earth's surface.

GF3. Carbon dioxide levels in the atmosphere have been increasing since the beginning of the industrial revolution.

IPCC/Gore's main hypotheses: "Anthropogenic forcing"

GH1. There is a greenhouse effect that warms the planet.

GH2. The greenhouse effect has the potential to be enhanced.

GH3. Global warming is a human-caused crisis.

Secondary hypotheses

GH4. Increasing the concentration of greenhouse gases in the atmosphere directly increases the warming of the Earth.

GH5. Small changes in global temperature have the potential to drastically upset a variety of climate systems through causal interactions.

Opposing hypotheses/beliefs

NH1. Long-term cycling of Earth's orbital parameters, solar activity, and volcanism and associated aerosols are natural causes that can warm the globe.

NH2. The impact of natural factors on global temperature dwarfs the enhanced greenhouse effect.

NH3. Climate systems will be affected by natural cycles and fluctuations.

NH4. Global warming is natural and not a concern.

SH1. Small changes in temperature will not have significant negative effects on global climate.

Explanations

Gore's explanations
  of the enhanced greenhouse effect and anthropogenic forcing:
    explain (GF1, GF2) GH1
    explain (GH1, GH4) GH2
    explain (GH2, GF3, GH5) GH3
  of the evidence:
    explain (GH2, GH3) E1
    explain (GH2, GH3) E2

explain (GH2, GH3, GH4, GF3) E3
explain (GH2, GH3, GH5) E4
explain (GH2, GH3, GH4, GF3) E5
Natural explanations
 of a natural cause for global warming:
 explain (NH1, NH2) NH4
 of the evidence:
 explain (NH1, NH2, NH4) E1
 explain (NH1, NH2, NH3, NH4) E4
Contradictions:
 contradict NH4 GH3
 contradict NH2 GH2
 contradict GH5 SH1

6 Coherence, Truth, and the Development of Scientific Knowledge

Introduction

The problem of the relation between coherence and truth is important for philosophy of science and for epistemology in general. Many epistemologists maintain that epistemic claims are justified, not by a priori or empirical foundations, but by assessing whether they are part of the most coherent account (see, e.g., Bonjour, 1985; Harman, 1986; Lehrer, 1990). A major issue for coherentist epistemology concerns whether we are ever warranted in concluding that the most coherent account is true. In the philosophy of science, the problem of coherence and truth is part of the ongoing controversy about scientific realism, the view that science aims at and to some extent succeeds in achieving true theories. The history of science is replete with highly coherent theories that have turned out to be false, which may suggest that coherence with empirical evidence is a poor guide to truth.

This chapter argues for a more optimistic conclusion, that coherence of the right kind leads to approximate truth. The right kind is explanatory coherence that involves theories that progressively broaden and deepen over time, where broadening is explanation of new facts and deepening is explanation of why the theory works. First, however, I will consider alternative accounts of the relation between coherence and truth, including the claims that coherence is truth, that coherence is irrelevant to truth, and that probability theory provides the link between coherence and truth.

I take coherence to be a relation among mental representations, including sentencelike propositions and also wordlike concepts and picturelike images. Coherence is a global relation among a whole set of representations, but it arises from relations of coherence and incoherence between pairs of representations. I describe below how this works in detail for explanatory coherence. As a preliminary account of truth, let me offer the following, adapted from Goldman (1999, p. 59): a representation such as a proposition is true if and only if it purports to describe reality and its content fits reality.

The theory I will develop about the relation of coherence and truth is naturalistic in two respects. First, my theory of coherence is psychologistic in that it employs a model of how human minds make inferences based on coherence considerations. Second, my main conclusion about how coherence can lead to truth is based on examples from the history of science, under the assumption that natural science is the major source of human knowledge. This chapter is not *about* naturalistic epistemology, but is an instance of it.

The Relation between Coherence and Truth

Before developing my own proposal for when coherence leads to truth, I shall deal very briefly with several other accounts of the relation of coherence to truth. The most audacious of these is the coherence theory of truth, according to which the truth of a representation consists in its coherence with other representations, not in its correspondence to a non-mental world; advocates include Blanshard (1939) and Rescher (1973). There are many standard objections to the coherence theory of truth (Young, 2008), but here I mention only a novel one. Coherence with scientific evidence strongly suggests that the universe is more than 10 billion years old, but that representations constructed by humans have existed for less than a million. Thus we can infer that there was a world existing independent of any human representation for billions of years. This inference does not in itself show that truth cannot consist in a relation only among representations, because a proponent of the coherence theory could simply maintain that there were no representations and hence no true representations until intelligent beings evolved. But if there is a world independent of representation of it, as historical evidence suggests, then the aim of representation should be to describe the world, not just to relate to other representations. My argument does not refute the coherence theory, but it shows that it implausibly gives minds too large a place in constituting truth.

Hence truth must consist of some sort of relation between the representations that occur in human minds or artifacts and the world. Truth is not a purely mental matter, because our best evidence suggests that minds and their representations have not been around all that long. The advocate of the coherence theory of truth could desperately contend that truth is coherence in the mind of God, but this supposes that the most coherent view includes belief in the existence of God, a supposition that I have challenged (Thagard, 2000, ch. 4). There are, of course, intensely skeptical challenges that can be made to this use of scientific evidence, but I will hold off addressing these until later.

A different way of trying to protect the coherence theory of truth against my argument would be to say that the bearers of truth are not mental representations but eternal abstract entities. On this view, propositions are not mental structures but Platonic objects constituting the meanings of sentences irrespective of whether there are any sentences. The problem with this Platonic reply is that we have no evidence that such abstract entities exist. In contrast, there is ample evidence from contemporary psychology and neuroscience that people employ mental representations, which therefore qualify as potential bearers of truth. Of course, the fact that, as far as we know, there were no mental representations 10 billion years ago does not undermine the correspondence theory of truth, because we can consider the fit, or lack of fit, between current representations and the state of reality at that time. The key point against the coherence theory of truth is that coherence with currently available evidence supports the view that reality is independent of representation of it.

At the other extreme from the coherence theory of truth, there is the view that coherence is simply irrelevant to truth. In epistemology, coherentist theories of knowledge are contrasted with foundational theories, which contend that knowledge is based not on a group of representations fitting with each other, but on a ground of indubitable truths. Rationalist foundationalists maintain that this ground consists of a set of truths known a priori, whereas empiricist foundationalists maintain that the ground consists of truths arising from sense experience. Unfortunately, both kinds of foundationalism have been dramatically unsuccessful in establishing a solid basis for knowledge. If there are any a priori truths, they are relatively trivial, such as Hilary Putnam's (1983) principle that not every statement is both true and false. No one has succeeded in constructing a priori principles that receive general agreement and enable the derivation of substantial knowledge (for further discussion, see Thagard, 2010a, ch. 2).

Similarly, the empiricist project of deriving knowledge from sense experience foundered because of the noncertain nature of sense experience and the nonderivability of scientific knowledge from experience alone. Our greatest epistemic achievements are scientific theories such as relativity theory, quantum theory, the atomic theory of matter, evolution by natural selection, genetics, and the germ theory of disease. None of these reduces to rationalist or empiricist foundations, so some kind of coherence theory of knowledge must be on the right track. (I am assuming that contextualism is a variety of coherentism, but see Henderson, 1994.) Rejection of a connection between coherence and truth is therefore tantamount to adopting general skepticism about the attainability of scientific knowledge. Later I will discuss skepticism arising from doubts about whether science really does attain truth.

I must also mention another prominent approach to relating coherence and truth that uses probability theory (e.g., Olsson, 2002; Shogenji, 1999). On this view, it should be possible to establish a connection between coherence and truth by means of an intermediary connection between coherence and probability. If we could show that propositions with greater coherence have higher probability, then we could judge that they are more likely to be true. This is a laudable project, but I see three insurmountable problems with it: interpretation, realization, and implementation.

The interpretation problem for probabilistic epistemology is the need to choose what meaning to assign to probabilities, which may be taken either as frequencies or degrees of belief. The frequency interpretation of probability clearly applies to scientific areas where data have been collected, but it does not apply at all to scientific theories. The probability of drawing a spade from a deck of cards is 0.25, meaning that in the long run the ratio of spades to cards drawn will approximate 0.25. But there are no comparable ratios applicable to scientific theories. On the other hand, viewing probabilities as degrees of belief is complicated by substantial evidence that human thinking does not naturally conform to Bayesian standards (Kahneman, Slovic & Tversky, 1982; Gilovich, Griffin & Kahneman, 2002). Normatively, one might insist that it should, but there would still be the gap between subjective degrees of belief and objective truth. Using probability as the connection between coherence and truth presupposes that there are links between (1) coherence and probability and (2) probability and truth. The frequentist interpretation of probability fails in making the first link, whereas the degree-of-belief interpretation fails in making the second link.

By the realization problem I mean the difficulty of analyzing coherence in such a way that these links can be made. Ingenious analyses of coherence in terms of probability have been made by Olsson (2002) and Shogenji (1999); but on their own terms they have been unsuccessful in connecting coherence and truth, independent of the problem of interpreting probability. A third problem, less commonly discussed in the philosophical literature, concerns the difficulty of implementing probabilistic reasoning computationally. Splendid computational tools have been developed for making inferences with probabilities (e.g., Pearl, 1988). But applying them to realistic cases of causal inference such as those involved in scientific or legal reasoning requires the concoction of large numbers of conditional probabilities that no reasoner would have available. For further discussion of the interpretation and implementation problems in the context of legal inference, see Thagard (2004).

Hence probability will not provide the needed connection between coherence and truth. Adherents of probabilistic epistemology would probably react by saying: so much the worse for coherence. But the intractability of the interpretation and implementation problems suggests a different response: so much the worse for probability, whose epistemological use is limited to areas like statistical inference, where the frequency interpretation applies. In the rest of this chapter, I will pursue a nonprobabilistic approach to the connection between coherence and truth.

Explanatory Coherence

This pursuit requires much greater specification of what coherence is, and for that purpose I will employ my theory of explanatory coherence whose principles were listed in chapter 5. These principles do not fully specify how to determine coherence-based acceptance, but algorithms are available that can compute acceptance and rejection of propositions on the basis of coherence relations. The most psychologically natural algorithms use artificial neural networks that represent propositions by artificial neurons or *units* and represent coherence and incoherence relations by excitatory and inhibitory links between the units that represent the propositions. Acceptance or rejection of a proposition is represented by the degree of activation of the unit. The ECHO program spreads activation among all units in a network until some units are activated and others are inactivated, in a way that maximizes the coherence of all the propositions represented by the units. ECHO has been applied to many cases of scientific and legal reasoning, without the implementation and interpretation problems that afflict probabilistic models of causal reasoning. The theory of explanatory coherence depends on the notion of explanation, which a later section discusses in terms of causal mechanisms.

The question now arises: do we have any reason to believe that hypotheses that are accepted as a group because they maximize explanatory coherence are at least approximately true? In ordinary life, counterexamples abound. For example, the theological and political beliefs of Osama bin Laden constituted a highly coherent set for him and his followers, but we would not want to acknowledge the truth of many of them. All of us have had the experience of making an inference to the best explanation about the behavior of a friend or the breakdown of a piece of machinery, only to learn that our inference was erroneous. Such everyday cases are often deficient, however, in considering the full range of evidence and alternative hypotheses, so perhaps if explanatory coherence had been assessed properly the erroneous inference might have been avoided. But the history of science contains many cases where theories high in explanatory coherence have turned out to be false.

The Pessimistic Induction

Newton-Smith (1981, p. 14) called the "pessimistic induction" the inference that any scientific theory will eventually be discovered to be false. Laudan (1981a) compiled a long list of theories from the history of science that support this induction, including:

- the crystalline spheres of ancient and medieval astronomy
- the humoral theory of medicine
- catastrophist geology

- the phlogiston theory of chemistry
- the caloric theory of heat
- the vital force theory of physiology
- the aether theories of electromagnetism and optics
- theories of spontaneous generation

The phlogiston theory, for example, had very substantial explanatory coherence, providing explanations of phenomena such as combustion and rusting that dominated chemistry for most of the eighteenth century. It was clearly the best explanation of the evidence until it was supplanted by Lavoisier's oxygen theory in the 1780s (Thagard, 1992, chs. 3–4). The pessimistic induction does not require that we know about the falsehood of previous theories by virtue of the truth of the theories that replaced them, which would make the induction incoherent. All it requires is noticing that many theories accepted as true were later rejected as false.

The pessimistic induction suggests that, since a great many theories in the history of science have turned out to be false, we should expect our current theories to turn out to be false as well. There may not be strong alternatives now for our most coherent theories such as relativity and evolution, but we should expect that eventually they will be superseded by theories with more explanatory coherence. Hence we should not associate the maximization of explanatory coherence with truth. The history of science thus suggests that coherence, even explanatory coherence along the lines I have suggested, is a poor guide to truth.

Various responses are available to the pessimistic induction (Psillos, 1999). It is noteworthy that Laudan's examples are all from before the twentieth century, and one could argue that recent science has been more successful in achieving true theories. After all, the personal and material resources of science have increased steadily over the past 100 years. However, the temporal induction that recent theories will turn out to be true seems rather shaky, because there just might not have been enough time for superior theories to come along and demonstrate the weaknesses of current theories. Smolin (2001) suggests that problems in making quantum theory and relativity theory compatible with each other may lead to the replacement of both by a quantum theory of gravity. We need a more epistemologically satisfying induction that can tell us when we can take a coherent theory to be true.

Whewell's Overoptimistic Induction

William Whewell, the great nineteenth-century historian and philosopher of science, identified a feature of theories that he thought identified ones that are true. He used the term "consilience of inductions" to describe the support for a hypothesis that comes when it enables us to explain cases of a kind different from those that were contemplated in its formation (Whewell, 1968, p. 153). Whewell had a comprehensive knowledge of the history of science to the mid-nineteenh century, and he generalized as follows: "No example can be pointed out, in the whole history of science, so far as I am aware, in which this Consilience of Inductions has given testimony in favour of an hypothesis afterwards discovered to be false" (pp. 154–155). Consilience requires a hypothesis to increase its explanatory coherence, not merely by explaining a new fact, but by explaining a kind of fact different from ones previously explained. If Whewell were right, then we would have the basis for an optimistic induction about the relation of coherence and truth: Theories that become more coherent over time by explaining new kinds of facts turn out to be true.

Whewell's favorite examples of consilience were Newton's theory of universal gravitation and the undulatory (wave) theory of light. A century and a half after Whewell's generalization, these examples appear rather unfortunate. Newton's theory of gravitation was never able to explain the perihelion of Mercury, but Einstein showed in 1915 that his theory of general relativity yielded more accurate predictions of it than did Newton (Gondhalekar, 2001). Thus, it appears that, contrary to Whewell's optimistic induction about a particular kind of coherence signaling truth, we have yet another instance of the pessimistic induction. Einstein rejected Newton's assumption that gravitational force is a function only of mass and distance, along with his assumptions of absolute space and time. Similarly, the wave theory of light has been superseded by the quantum theory that views light as consisting of particle-like photons that have wavelike properties.

An even more damning counterexample to Whewell's optimistic induction is the phlogiston theory, which was originally developed to explain combustion by Johann Becher, who called the principle of inflammability *terra pinguis*. Becher and Georg Stahl, who renamed this substance "phlogiston," applied it also to explain calcination (rusting) and respiration.

Since rusting and breathing appear very different from burning, extension of the phlogiston theory to calcination seems to constitute an instance of the consilience of inductions; but already by Whewell's time the falsity of the phlogiston theory had been recognized (Partington, 1961).

Thus, Whewell was overoptimistic about the epistemic power of the consilience of inductions. Still, I think he was on the right track in looking for temporal properties of developing theories that might mark them as good candidates for truth. At least we can say that the theory of universal gravitation and the wave theory of light are not so totally false as the theories of crystalline spheres, phlogiston, and caloric turned out to be. Below I will try to identify a sense in which Newtonian gravitation is partly true, but first I want to discuss a truth-related mark of coherent theories that is more promising than consilience.

Deepening and the Cautiously Optimistic Induction

There are two main ways in which a hypothesis can increase its explanatory coherence over time. The first is to explain new facts, which I will call "broadening." Whewell's consilience is a special kind of broadening in which the new facts explained are of a kind different from those already explained. The consilience of Wilson (1998) is an even more special kind of broadening that involves the interlocking of causal explanations across disciplines.

The second way in which a hypothesis can increase its explanatory coherence is by being explained by another hypothesis, which I will call "deepening." The process of a hypothesis being explained is most easily understood by legal examples in which questions of motive arise. The hypothesis that an accused is responsible for killing a victim gets its main explanatory coherence from its ability to explain a range of evidence, for example, that the accused's fingerprints are on the murder weapon. This hypothesis can be broadened by finding new evidence that it explains, for example, that one of the victim's hairs was found on the accused's clothes. Homicide detectives also want to know about possible motives of the accused, in order to provide an explanation of why the accused committed the murder. For example, the case that O.J. Simpson killed his ex-wife was based mainly on evidence that he did so, but also on his having the motive of jealousy for attacking her and her boyfriend. We then get a deeper explanation of the crime, because the motive explains the killing, which explains the evidence. In the normal practice of the law, the explanation draws on folk rather than on scientific psychology, attributing the behavior of the accused to ordinary beliefs, desires, and emotions rather than to richer structures and processes for which there is experimental evidence.

Normally, the deeper hypothesis is not adduced merely for the purpose of explaining the basic one, but has independent evidence supporting it. Figure 6.1 shows the structure of a hypothesis A that has broadened by explaining pieces of evidence 2 and 3 in addition to evidence 1, and has deepened by being explained by hypothesis B.

In science, deepening occurs when an explanation provides an underlying causal basis for a causal hypothesis. For example, consider the germ theory of disease, which says that contagious diseases are caused by microorganisms such as bacteria, viruses, and fungi. (To modern ears, this might sound like a tautology, but contagion was recognized long before the pathological effects of germs.) A particular instantiation of the germ theory identifies a specific disease that is caused by a specific microbe; for example, influenza is caused by viruses of the family *Orthomyxoviridae*. This theory has been deepened over the years by microbiological accounts that explain how viruses infect cells, replicate themselves, and disrupt the functions of cells and organs. For many viruses, biologists have identified all of their genes and the functions they perform. Thus we know not only that myxoviruses cause different types of influenza, but how they do so by their mechanisms of attachment, infection, and reproduction. Some other

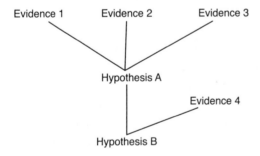

Figure 6.1
Hypothesis A is broadened by explaining new kinds of evidence and deepened by being explained by hypothesis B that explains additional evidence.

examples of deepening in recent history of science include the use of microbiology to explain how genetic transmission works and the use of the quantum-mechanical theory of molecular bonding to explain how atoms combine into molecules, which explains molecular theories of chemical reactions.

Now let me offer my own version of an optimistic induction about the relation between coherence and truth, which I call the "deepening maxim":

Explanatory coherence leads to truth when a theory not only is the best explanation of the evidence, but also broadens its evidence base over time and is deepened by explanations of why the theory works.

I am unaware of any broadened and deepened theory that turned out to be mostly false, but am aware that counterexamples may arise, so let me call this the "cautiously" optimistic induction. Actually, we do not need a universal generalization here: it would be enough if we could show from a survey of the history of science that broadened and deepened theories rarely turn out to be false. It is remarkable that none of the theories that I discussed in connection with the pessimistic induction were ever deepened. That is, no underlying mechanisms were identified for how entities such as phlogiston and caloric worked. The deepening maxim is a generalization not only about past theories but about future ones, predicting that future theories that are broadened and deepened will tend to be true.

My response to the pessimistic induction is very different from one recently criticized by Stanford (2003). He argues that scientific realism is undermined rather than supported by attempts to show that discarded theories such as phlogiston and caloric were at least partially successful with respect to the reference of central terms, core causal descriptions, partial truth, select preservation, and/or historical continuity. Instead of trying to defeat the pessimistic induction by arguing that the discarded theories are at least partly true, my strategy is to admit their falsehood and look for features that mark current theories as promising candidates to avoid joining phlogiston and caloric in the dustbin of history. The combination of broadening and deepening seems to be such a feature, but more needs to be said about how mechanistic explanations are deepened.

The importance of broadening and deepening is entailed by the theory of explanatory coherence discussed above. It follows from principle E2,

Explanation, that the more a hypothesis explains, the more coherent it is, other things being equal. Hence finding a new fact that is explained by a theory increases its explanatory coherence. Principle E2 also implies that a hypothesis explained by another hypothesis is more coherent, other things being equal. The theory of explanatory coherence by itself is neutral about the nature of explanation, however, but it fits well with the view that explanations are based on causal mechanisms.

Mechanisms and Explanation

To explicate the deepening maxim further, I need to say much more about how a theory can receive a deeper explanation. This requires an account of the nature of explanation, which can be provided by attention to the nature of mechanisms. It then becomes possible to characterize the nature of approximate truth in terms of the ontology of mechanisms.

In accord with much recent work in the philosophy of science, I hold that to explain a phenomenon is to describe a mechanism that produces it (see the discussion of the mechanista approach in chapter 1). A mechanism is a system of parts whose properties and relations produce regular changes in those properties and relations. For example, we explain how brains work by specifying their parts, which are neurons organized into neuronal groups and functional areas. Neurons have properties such as their electric potentials, and relations such as their links to other neurons via excitatory and inhibitory synapses. We can use mathematical equations and computer models to infer the behavior of groups of neurons from their properties and relations, including inputs from an external environment.

If explanation is mechanism based, we can develop an account of deepening in terms of parts. A deeper explanation for an explanatory mechanism M1 is a more fundamental mechanism M2 that explains how and why M1 works. M1 consists of parts, and M2 describes parts of those parts whose properties and relations change in ways that generate the changes in the properties and relations of the parts in M1. For example, neural mechanisms are deepened by the fact that neurons consist of parts— proteins and other molecules that are organized into functional areas such as the nucleus, mitochondria, axons, dendrites, and synapses. Chemical reactions involving these proteins enable nerve cells to carry out their basic functions, including taking inputs from presynaptic neurons, building up electric charges, spiking, and sending signals to postsynaptic neurons. Thus, neuroscience is deepened by molecular biochemistry, which explains how neurons work. This is a special case of how cell biology is deepened by molecular biochemistry, as has progressively happened since the 1960s. Similarly, medical theories such as the germ theory of disease have been deepened by finding lower-level mechanisms for the operations of cells and their microbial invaders. As I mentioned in the previous section, medicine knows enough about the parts of bacteria, viruses, and fungi to be able to explain how they invade and disrupt the normal function of bodily cells.

Deepening is also pervasive in modern chemistry and physics. We explain the chemical behavior of elements and compounds by the atoms that compose them. In turn, contrary to the original ancient and nine-teenth-century atomic theories, the behavior of atoms can be explained by describing their parts, right down to the level of quarks and possibly superstrings. I am not assuming the traditional philosophical view of reductionism, according to which the deeper theory serves to deductively predict what goes on at the higher level: the complexity and sensitivity to chaos of higher- and lower-level systems may make such predictions impossible. But we gain much understanding nevertheless by noting that the mechanism at the upper level works as it does because of the operations of the parts at the lower level. The deepening maxim obviously does not apply to the most fundamental level in subatomic physics, but still has ample room for application in other areas of physics as well as chemistry, biology, and the social sciences. The instances claimed to support the pessimistic induction are not at the fundamental level at which deepening does not apply.

Thus, in important areas of medicine, biology, chemistry, and physics, we commonly get deepening by theoretical mechanisms that show how the parts, properties, relations, and changes at the higher level decompose into parts at the lower level. Evidence for the lower-level mechanism is not just that it explains the higher-level one. For example, there is ample molecular and chemical evidence for the structure and operations of neurons and the operations of microbes, so molecular biochemistry provides support for the acceptability of the neuronal theory of brain operation and for the germ theory of disease. The deepening maxim can then be specified as the induction that theories can be judged to be true if they have been deepened by having the mechanisms they describe decomposed into more fundamental mechanisms for which there is independent evidence. As we have seen, inductive support for the deepening maxim includes the germ theory of disease, the neuronal theory of brains, molecular cell biology, molecular genetics, and the atomic theory of matter. Attention to mechanistic explanation serves to spell out and support my cautiously optimistic induction about the connection between coherence and truth.

Approximate Truth

However, some cases from the history of science may constitute challenges to the deepening maxim and require weakening it to conjecture a connection only between coherence and approximate rather than absolute truth. Consider, for example, the atomic theory of matter. I used this as an example of deepening: we can now explain how atoms undergo changes such as forming molecules by using quantum-mechanical theories about their parts. But I also noted that this deepening required abandonment of the previously definitional truth that atoms do not have parts. So, strictly speaking, the atomic theory was not deepened, but rendered false. It would be a mistake, however, to treat a theory as merely a conjunction of hypotheses that is false if any one of them is false. Instead, we should spell out what the theory claims about mechanisms consisting of parts, properties, relations, and resultant changes, and identify how many of these claims turn out to be wrong. We can then maintain that the atomic theory of matter has survived because most of its claims about the constituents of things are still taken to be true: elements and compounds have properties, relations, and changes that result from the atoms of which they consist.

Newtonian mechanics is another difficult case. I used it to challenge as overoptimistic Whewell's induction about consilience. But it is also a possible challenge to my deepening maxim: it might be argued that general relativity deepens Newtonian theory by providing an explanation in terms of the curvature of space-time of how Newton's force of gravity works. Then Newton's theory of gravitation seems like a counterexample to the deepening maxim, in that it has been deepened but is acknowledged to be false by virtue of its failed predictions and rejected assumptions such as that mass is independent of energy.

However, to say that general relativity totally replaced Newtonian gravitation would be as much of a mistake as saying that it fully incorporates it (see Thagard, 1992, pp. 214–215, for details). Newton's three laws and the principle of gravitation are good approximations to what general relativity predicts as long as velocities and masses are small. Although Newtonian mechanics does not accurately predict the perihelion of Mercury, its degree of inaccuracy was only 8 percent: 43 arcseconds per century (Gondhalekar, 2001, p. 162). So it is reasonable to maintain that Newtonian mechanics is approximately true in the sense that its major claims are quantitatively close to those supported by evidence and the theory that replaced it. I take a theory to be approximately true if it is *partly* true, that is, if *most* if its claims are *nearly* true in achieving quantitative closeness to accepted values. Assessment of approximate truth does not simply involve counting sentences, but needs to qualitatively consider the central mechanistic claims that the theory makes about parts, properties, relations, and resulting changes. For further discussion of approximate truth, also known as verisimilitude and truthlikeness, see Psillos (1999), Kuipers (2000), and Thagard (2000).

My cautiously optimistic induction is thus cautious in two respects. First, it allows for the possibility that a major instance of a deepened theory could turn out to be false. I do not expect fields such as molecular medicine, genetics, and atomic theory to be radically overthrown, but it could happen. Second, it allows for the possibility, which seems to have happened in both the atomic theory of matter and Newtonian mechanics, that deepening by virtue of a more fundamental mechanism may lead to some revisions in the original theory, with recognition that it is only approximately (partly and nearly) true. Accordingly, here is my final version of the deepening maxim: If a theory not only maximizes explanatory coherence, but also broadens its evidence base over time *and* is deepened by explanations of why the theory's proposed mechanism works, then we can reasonably conclude that the theory is at least approximately true. This induction is the strongest relation available between coherence and truth.

If a theory is broadened and deepened, is it still the same theory? As in the discussion of approximate truth, it is useful to think of a theory not as just a set of sentences but rather as a representation of parts, properties, relations, and changes. Broadening a theory by finding a new explanatory application of it and deepening it by identifying an underlying mechanism clearly do not generate a new theory, as long as the original hypotheses about parts, properties, relations, and changes remain substantially the same.

Deepening the Deepening Maxim

The deepening maxim gains credibility from the numerous theories in the history of science, such as the germ theory of disease and the atomic theory of matter, that have both undergone deepening and avoided the dustbin of rejected theories. But it would gain increased credibility if we could say *why* it is true that deepened theories tend to hold up well to empirical investigation and therefore appear to be approximately true. Deepening is not a necessary condition of the acceptability of a scientific or philosophical claim, because knowledge about underlying mechanisms may simply not be available at a given time. The germ theory of disease, the theory of evolution by natural selection, and atomic theory were all credible before microbiology, genetics, and subatomic physics were developed. Nevertheless, they gained additional credibility when underlying mechanisms became understood.

Hence, it is useful to ask why deepened theories tend to avoid becoming instances that support the pessimistic induction. One superficial explanation might be that deepened theories tend to survive because scientists simply prefer deepened theories as part of their overall strategy of maximizing explanatory coherence. Deepened theories survive not because they are good candidates for approximate truth, but merely because they are popular with scientists. The flaw in this explanation is that it ought also to apply to broad theories: broad theories survive just because they are what scientists like, not because they have anything to do with truth. But we have seen abundant examples of broad theories that were superseded by broader theories, as oxygen superseded phlogiston and thermodynamics replaced caloric. Even theories that underwent broadening—recall Whewell's examples of Newtonian mechanics and the wave theory of light—have given way to successors.

Following a suggestion of Peter Railton's (personal communication), I think that the most plausible explanation of why deepened theories survive and thrive is that they are at least approximately true. The success of theories for whom underlying mechanisms are found is not the result of scientific fashion, but rather the result of a strategy that fits well with the structure of the world that science investigates. Different kinds of experiments and instruments enable science to study the structure of the world at different levels, and these levels fit together naturally. For example, when a health researcher can transfer illness from one animal to another, the tools of bacteriology and virology explain why this is so. In turn, the tools of molecular biology enable an explanation of how viruses and bacteria infect cells by means of molecular signals. In turn, the tools of physical chemistry enable an explanation of how atoms form molecules and how chemical reactions occur. Here the deepening strategy works because the world is in fact organized in terms of parts, from organisms down to subatomic particles, and layers of mechanisms, from viral infection down to chemical bonding. Hence, the best explanation of the great success of deepened theories is that they are finding out about the world at multiple levels. Scientific realism should not consider individual theories in isolation from each other, but should notice how well the most successful of them manage to nest vertically with each other as well as to fit with experimental observations. This nesting, accomplished by instruments and experiments and theorizing that operate at multiple interacting levels, is the key to a deep account of the relation between coherence, deepening, and truth.

The argument in this section is a special case of the general abductive argument for scientific realism, which has the form:

(1) Science is successful with respect to predictions, technological applications, cumulation of knowledge, and agreement among practitioners.

(2) The best explanation of this success is that scientific theories are at least approximately true.

(3) Therefore, scientific theories are at least approximately true.

This is not the place to discuss or defend the general argument (see Thagard, 1988, ch. 8; Kitcher, 2002). I am presenting a more specific argument:

(4) Deepening maxim: Theories that have undergone deepening by lower-level mechanisms have survived empirical investigation.

(5) The best explanation of this survival is that scientific theories about layers of mechanisms are at least approximately true.

(6)   Therefore, scientific theories about layers of mechanisms are at least approximately true.

Notice that (6) follows deductively from (3), so that the general support for scientific realism supports scientific realism about mechanisms, which then supports the deepening maxim. The support is mutual, as is generally the case in explanatory coherence: When a hypothesis explains evidence or another hypothesis, both the explainer and the explained support one another. In the current context, the crucial point is that the deepening maxim has been deepened by a hypothesis about the approximate truth of theories about layers of mechanisms.

Levin (1984) challenged scientific realists to give *mechanistic* explanations of how truth produces success, which would require at least a sketch of the mechanisms by which the truth of theories about layers of mechanisms produces the success described by the deepening maxim. This task is daunting, but should be doable for particular cases such as my medical example relating infectious diseases to underlying mechanisms. The relevant mechanisms include use of instruments that detect features of the world at different levels of detail, ranging from stethoscopes to MRI machines to optical and electron microscopes. Defending the reliability of these instruments provides further support for scientific realism, in accord with the "Galilean strategy" of Kitcher (2001). Additional relevant mechanisms include physical and social processes by which experiments are carried out, and cognitive mechanisms by which human minds collect and interpret data. Thus filling out the explanation in (5) that links approximate truth with deepening is very complex, but there is reason to believe that it could be done mechanistically.

Conclusion

I have argued that many of the standard claims about the relation of coherence to truth—that coherence is truth, that coherence is irrelevant to truth, and that coherence leads to probability which leads to truth—are implausible. Instead, I have built on my theory of explanatory coherence and the mechanism-based theory of explanation deepening to produce a cautiously optimistic induction about when coherence usually leads to truth in natural science. This induction constitutes a response to doubts about whether coherence connects to truth. (Millgram, 2000, raises the question of whether *approximation* to coherence of the sort performed by available algorithms leads to truth; for a response, see Thagard, forthcoming-a.)

To conclude, I will discuss the implications of this view of the relation between coherence and truth for social science, everyday life, and philosophical deliberation about the nature of knowledge. Social science has not witnessed the extent of deepening found in the natural sciences, but there are some good prospects. Cognitive psychology is now heavily enmeshed with neuroscience, and cognitive theories, which were previously couched only in terms of representations and computations, are increasingly being fleshed out in terms of neural structures and processes. The neurological turn in cognitive psychology, and to a lesser extent in other areas of psychology such as social, developmental, and clinical psychology, opens the possibility of the field being deepened by neuroscience, just as neuroscience is being progressively deepened by neurochemistry. Another promising trend is that economics, which formerly dealt with highly idealized models of human rationality that had little to do with human psychology, is increasingly tied with cognitive psychology and neuroscience, through the development of the field of behavioral economics. I am not raising the prospect of reducing economics to psychology and psychology to neuroscience, but rather pointing to the salutary trend of economics enriching its theories with information about psychological mechanisms, and psychology enriching its understanding with information about neurological mechanisms, which in turn are deepened by biochemical mechanisms. Hence, there is a reasonable prospect that my cautiously optimistic induction may eventually apply to social science as well as natural science.

What about everyday life? Most epistemologists have worried about people's ordinary knowledge, such as what we are purported to know about the external world and other people. Although the deepening maxim does not apply to most beliefs of most people, this does not mean that they are grossly unwarranted. The deepening maxim specifies conditions under which coherence very reliably leads to truth, but it does not imply that there is no truth without deepening. Broad explanatory theories in everyday life may well turn out to be often true, for example, our common, everyday beliefs about physical objects and the behavior of other people.

More importantly, if we step outside everyday beliefs and detect their origins in psychophysical processes, we can deepen everyday explanations considerably. For example, beliefs about physical objects that arise from sense perception can be understood more deeply by going beyond ordinary people's knowledge about seeing, through appreciation of the physical and neurological processes that connect objects and our perception of them by photons of light, retinal stimulation, and neural processing. Physics and neuroscience then provide mechanisms explaining why people have experiences and beliefs about the external world, thereby deepening their knowledge about it. Similarly, scientific psychology and neuroscience provide a basis for explaining why folk psychology is sometimes right and sometimes wrong about why people think and behave the ways they do.

Whereas commonsense perception and cognition are open to deepening by physical and neurological mechanisms, skeptical theories are not. Skeptics have suggested that our experiences may arise not from veridical psychological processes, but by the machinations of a deceptive god, evil genius, or matrix of brains in vats. These alternative theories are inherently shallow, in that there are no evidence-supported mechanisms that explain how we could be so systematically deceived. To consider only the most recent example, the hypothesis from *The Matrix* movies, that human experiences arise from illusory inputs from intelligent machines, fails to take into account the physical and computational implausibility of generating the complexity and rapidity of such experiences. It takes years to produce an animated movie such as *Finding Nemo*, so there is no plausible technology that could produce the multiple streams of interconnected experiences portrayed in *The Matrix*. Thus commonsense epistemology is open to deepening by reasonable extensions of current physics, psychology, and neuroscience, whereas skeptical epistemologies float flimsily in the air. Hence, developments in the cognitive sciences should eventually have the result that the everyday hypothesis that our perceptions arise largely as the result of interactions with an external world will fall within the scope of the cautiously optimistic induction. Then naturalistic epistemology will mesh with the philosophy of science to provide a deep justification of everyday as well as scientific knowledge (see Thagard, 2010a, ch. 4).

III Discovery and Creativity

7 Why Discovery Matters

How do scientists make discoveries? This question is important not only for appreciating scientific investigations that are currently underway, but also for understanding great discoveries of the past. Historians, philosophers, and psychologists have wondered about how Newton came up with his theory of gravitation, how Darwin generated his theory of evolution by natural selection, and how Einstein produced the theory of relativity. Cognitive science can contribute to answering these questions by describing the mental structures and processes that enable scientists to make the full range of scientific discoveries, including observing new phenomena, generating new concepts, forming new explanatory hypotheses, and developing new methods and procedures.

There are several reasons why discovery is an important topic for cognitive science. First, discovery is the most exciting part of science, both for scientists who make the breakthroughs and for enthusiastic spectators. As I chronicled in an essay "The Passionate Scientist," the thrill of discovery is a major part of the motivation that impels scientists to labor intensively toward sometimes elusive goals (Thagard, 2006, ch. 10). Hence, we cannot understand the practice of science without appreciating how discoveries arise.

Second, discovery is relevant to issues about explanation and justification discussed in Part II. When scientists evaluate competing theories to determine which hypotheses provide the best explanation of the evidence, they do not attempt to evaluate all possible theories, which would be cognitively impossible. Rather, they consider the available alternatives, often only two but rarely more than a few. Generating plausible new theories along with the concepts and methods that go with them is not easy. Hence the discovery process puts a kind of filter on the availability of hypotheses subject to further evaluation. Unlike biological mutation, which generates variants unconnected with their likelihood of success, the process of scientific discovery is highly intentional and produces a more focused set of hypotheses that can then be the subject of evaluation with respect to their explanatory coherence.

Third, the understanding of discovery is relevant to science education because of the need to motivate students to acquire new concepts, theories, and methods. Motivation should be increased if students are not simply force-fed a stock of information to acquire, but can also get some sense of the thrill of figuring things out for themselves. Appreciating the cognitive and social mechanisms that produce discoveries may lead to social policies that encourage education programs that increase scientific and technological productivity.

Fourth, the study of scientific discovery is part of the larger enterprise of investigating creativity in general. Cognitive science needs to be concerned with everyday cognition, involving frequently used processes of problem solving, learning, and so on. But it should also be able to explain the much rarer creative leaps that produce important breakthroughs in science, technology, the arts, and social institutions. It is an interesting question whether the same cognitive processes are involved in the creative leaps in scientific discovery, technological invention, artistic imagination, and social innovation.

Scientific discovery was of interest to early philosophers of science such as Bacon, Whewell, Mill, and Peirce. In the 1900s, however, logical positivists argued that philosophy should only be concerned with the justification of scientific theories, leaving issues about discovery to psychologists. N. R. Hanson (1958) was a philosopher who revived interest in discovery, amplified by increased attention to the history of science inspired by the work of Kuhn (1962) and others. There followed an upswing in interest in discovery as a philosophical problem (Nickles, 1980). This trend was amplified when philosophers such as Lindley Darden began to take an interest in the emerging practice of computational modeling of discovery (Darden, 1983).

Herbert Simon was one of the founders of artificial intelligence and cognitive science in general, beginning with his pathbreaking work on human problem solving from a computational perspective (Newell, Shaw & Simon, 1958). In the 1980s, he collaborated with Pat Langley and others to produce computational models of discoveries such as Kepler's laws of planetary motions (Langley et al., 1987). Much subsequent work by computer scientists has looked at processes by which computers and people can generate new hypotheses (e.g., Bridewell et al., 2008; Bridewell & Langley, 2010; Kulkarni & Simon, 1988, 1990; Valdes-Perez, 1995). There is even a Robot Scientist that tests hypotheses as well as generating them (King et al., 2009).

The main methods of psychologists—behavioral experiments—do not enable them directly to study scientists in the process of discovery, but psychologists have devised many ways of examining the cognitive processes that underlie discovery. Relevant psychological methods include experiments to study complex problem solving, insight (Ohlsson, 2011; Sternberg & Davidson, 1995), and analogy (Holyoak & Thagard, 1995). Kevin Dunbar has been particularly ingenious in studying high-level cognition by using multiple techniques ranging from video taping lab meetings to scanning the brains of problem solvers (Dunbar & Fugelsang, 2005). Experimental work has just begun on the neuroscience of discovery, with interesting studies of the neural correlates of insight (Kounios & Beeman, 2009; Subranamiam et al., 2009). Useful collections of work on the psychology of discovery include Ward, Smith, and Vaid (1997) and Finke, Ward, and Smith (1992). On creativity from both psychological and computational perspectives, see Boden (2004). Other books on the psychology of creativity include Perkins (2001) and Weisberg (1993).

I know of little research in cognitive anthropology or linguistics on discovery processes of scientific cognition. Latour and Woolgar (1986) approached laboratory life from an anthropological perspective, but repudiated cognitive explanations of science. Nersessian (2009) presents an ethnographic study of conceptual innovation in an engineering laboratory. As I stressed in the introduction, I see no opposition between social explanations of science and cognitive ones: like the rest of science, discovery is both a social and a cognitive process. See also Osbeck et al. (2011).

It should be clear from the above quick survey that cognitive science has contributed in many ways to understanding scientific discovery through research in philosophy, psychology, and artificial intelligence. Although there is no definitive cognitive theory of discovery, many of the relevant pieces have been identified, including cognitive processes for generating new ideas and hypotheses. The chapters in Part III are some of my own recent work on discovery and creativity. Chapter 8 provides a neurocomputational account of how creative conceptual combination works in the brain. It assumes the *combinatorial conjecture* that new ideas arise from the combination of old ones, and chapter 9 provides a test of this conjecture by looking at a large number of examples of scientific discoveries and technological inventions. It provides support for the claim defended by Thagard and Croft (1999) that creative breakthroughs in science and in technology use the same basic set of cognitive processes, although it also finds some interesting differences such as the greater role of serendipity in science and the greater role of visual representations in technology. Chapters 10 and 11 furnish reviews of discovery in more specific areas—computer science and medicine.

One important neglected aspect of discovery might be called "procedural creativity," by which I mean the discovery or invention of new methods. Creativity in art is often the result of devising new techniques, such as the stream of consciousness writing pioneered by James Joyce and the painting with small brush strokes used by the impressionists. Similarly, a big part of scientific development is not just the generation of new concepts and hypotheses, but also the introduction of new methods, such as Galileo's use of the telescope to examine the skies, Hooke's use of the microscope to study plants, and the twentieth-century invention of polymerase chain reaction that made possible detailed studies of DNA. I would like to see more historical, psychological, and computational investigations of how people create new methods. The conceptual combinations described in chapter 8 will probably be part of the story, but much more of it remains to be figured out. There also needs to be a systematic comparison of creativity in scientific discovery, technological invention, artistic imagination, and social innovation.

8 The *Aha!* Experience: Creativity through Emergent Binding in Neural Networks

Paul Thagard and Terrence C. Stewart

Creative Cognition

Creativity is evident in many human activities that generate new and useful ideas, including scientific discovery, technological invention, social innovation, and artistic imagination. Understanding is still lacking of the cognitive mechanisms that enable people to be creative, especially about the neural mechanisms that support creativity in the brain. How do people's brains come up with new ideas, theories, technologies, organizations, and aesthetic accomplishments? What neural processes underlie the wonderful *Aha!* experiences that creative people sometimes enjoy?

We propose that human creativity requires the combination of previously unconnected mental representations constituted by patterns of neural activity. Then creative thinking is a matter of combining neural patterns into ones that are both novel and useful. We advocate the hypothesis that such combinations arise from mechanisms that bind together neural patterns by a process of convolution rather than synchronization, which is the currently favored way of understanding binding in neural networks. We describe computer simulations that show the feasibility of using convolution to produce emergent patterns of neural activity of the sort that can support human creativity.

One of the advantages of thinking of creativity in terms of neural representations is that they are not limited to the sort of verbal and mathematical representations that have been used in most computational, psychological, and philosophical models of scientific discovery. In addition to words and other linguistic structures, the creative mind can employ a full range of sensory modalities derived from sight, hearing, touch, smell, taste, and motor control. Creative thought also has vital emotional components, including the reaction of pleasure that accompanies novel combinations in the treasured *Aha!* experience. The generation of new representations involves binding together previously unconnected representations in ways that also generate new emotional bindings.

Before getting into neurocomputational details, we illustrate the claim that creative thinking consists in novel combinations of representations with examples from science, technology, social innovation, and art. We then show how multimodal representations can be combined by binding in neural populations using a process of convolution in which neural activity is "twisted together" rather than synchronized. Emotional reactions to novel combinations can also involve convolution of patterns of neural activity. After comparing our neural theory of creativity with related work in cognitive science, we place it in the context of a broader account of multilevel mechanisms—including molecular, psychological, and social ones—that together contribute to human creativity.

We propose a theory of creativity encapsulated in the following theses:

1. Creativity results from novel combinations of representations.

2. In humans, mental representations are patterns of neural activity.

3. Neural representations are multimodal, encompassing information that can be visual, auditory, tactile, olfactory, gustatory, kinesthetic, and emotional, as well as verbal.

4. Neural representations are combined by convolution, a kind of twisting together of existing representations.

5. The causes of creative activity reside not just in psychological and neural mechanisms, but also in social and molecular mechanisms.

Thesis 1, that creativity results from combination of representations, has been proposed by many writers, including Koestler (1967) and Boden (2004). Creative thinkers such as Einstein, Coleridge, and Poincaré have described their insights as resulting from combinatory play (Mednick, 1962). For the purposes of this chapter, the thesis need only be the modest claim that *much* creativity results from novel combination of representations. The stronger claim that *all* creativity requires novel combination of representations is defended by an analysis of 200 great scientific discoveries and technological inventions (chapter 9). The nontriviality of the claim that creativity results from combination of representations is shown by proponents of behaviorism and radical embodiment who contend that there are no mental representations.

Thesis 2, that human mental representations are patterns of neural activity, is defended at length elsewhere (Thagard, 2010a). The major thrust of this chapter is to develop and defend thesis 4 by providing an account of how patterns of neural activity can be combined to constitute new ones. Our explanation of creativity in terms of neural combinations could be taken as an additional piece of evidence that thesis 2 is true, but we need to begin by providing some examples that provide anecdotal evidence for theses 1 and 3. Thesis 5, concerning the social and molecular causes of creativity, will be discussed only briefly.

Creativity from Combination of Representations

Fully defending thesis 1, that creativity results from combination of representations, would take a comprehensive survey of hundreds or thousands of acknowledged instances of creative activity in many domains. This chapter provides a couple of supporting examples from each of the four primary areas of human creativity: scientific discovery, technological invention, social innovation, and artistic imagination. These examples show that at least some important instances of creativity depend on the combination of representations.

Many scientific discoveries can be understood as instances of conceptual combination, in which new theoretical concepts arise by putting together old ones (Thagard, 1988). Two famous examples are the wave theory of sound, which required development of the novel concept of a sound wave, and Darwin's theory of evolution, which required development of the novel concept of natural selection. The concepts of sound and wave are part of everyday thinking concerning phenomena such as voice and water waves. The ancient Greek Chrysippus put them together in to create the novel representation of a sound wave that could explain many properties of sound such as propagation and echoing. Similarly, Darwin combined familiar ideas about selection done by breeders with the natural process of struggle for survival among animals to generate the mechanism of natural selection that could explain how species evolve.

One of the cognitive mechanisms of discovery is analogy, which requires putting together the representation of a target problem with the representation of a source (base) problem that furnishes a solution. Hence, the many examples of scientific discoveries arising from analogy support the claim that creativity arises from combination of representations. See Holyoak and Thagard (1995) for a long list of analogy's greatest successes in the field of scientific discovery.

Cognitive theories of conceptual combination have largely been restricted to verbal representations (e.g., Smith & Osherson, 1984; Costello & Keane, 2000), but conceptual combination can also involve perception (Wu & Barsalou, 2009; see also Barsalou et al., 2003). Obviously, the human concept of sound is not entirely verbal, possessing auditory exemplars such as music, thunder, and animal noises. Similarly, the concept of wave is not purely verbal, but involves in part visual representations of typical waves such as those in large bodies of water or even in smaller ones such as bathtubs. Hence, a theory of conceptual combination, in general and in specific application to scientific discovery, needs to attend to nonverbal modalities.

Technological invention also has many examples of creativity arising from combination of representations. In the authors' home town of Waterloo, Ontario, the major economic development of the past decade has been the dramatic rise of the company Research in Motion (RIM), maker of the extremely successful BlackBerry wireless device. The idea for this device originated in the 1990s as the result of the combination of two familiar technological concepts: electronic mail and wireless communication. According to Sweeny (2009), RIM did not originate this combination, which came from a Swedish company, Ericsson, where an executive combined the concepts *wireless* and *email* into the concept *wireless email*. Whereas the concept of sound wave was formed to explain observed phenomena, the concept of wireless email was generated to provide a new target for technological development and financial success (see chapter 10 for an account of creativity in computer science). Thus, creative conceptual combination can produce representations of goals as well as of theoretical entities. RIM's development of the BlackBerry depended on many subsequent creative combinations such as two-way paging, thumb-based typing, and an integrated single mailbox.

Another case of technological development by conceptual combination is the invention of the stethoscope, which came about by an analogical discovery in 1816 by a French physician, Théophile Laennec (Thagard,

1999, ch. 9). Unable to place his ear directly on the chest of a modest young woman with heart problems, Laennec happened to see some children listening to a pin scratching through a piece of wood, and came up with the idea of a *hearing tube* that he could place on the patient's chest. The original concepts here are multimodal, involving sound (hearing heartbeats) and vision (rolled tube). Putting these multimodal representations together enabled Laennec to create what we now call the stethoscope. It would be easy to document dozens of other examples of technological invention by representation combination.

Social innovations have been less investigated by historians and cognitive scientists than developments in science and technology (Mumford, 2002), but these also result from representation combination. Two of the most important social innovations in human history are public education and universal health care. Both of these innovations required establishing new goals using existing concepts. Education and health care were private enterprises before social innovators projected the advantages for human welfare if the state took them on as a responsibility. Both innovations required novel combinations of existing concepts concerning government activity plus private concerns, generating the combined concepts *public education* and *universal health care*. Many other social innovations, from universities, to public sanitation, to Facebook, can also be seen as resulting from the creative establishment of goals through combinations of previously existing representations. The causes of social innovation are, of course, social as well as psychological, as we will make clear later when we propose a multilevel system view of creativity.

There are many kinds of artistic creativity, in domains as varied as literature, music, painting, sculpture, and dance. Individual creative works such as Beethoven's *Ninth Symphony*, Tolstoy's *War and Peace*, and Manet's *Le déjeuner sur l'herbe* are clearly the result of the cognitive efforts of composers, authors, and artists to combine many kinds of representations: verbal, auditory, visual, and so on. Beethoven's *Ninth*, for example, combines auditory originality with verbal novelty in the famous last movement known as the *Ode to Joy*, which also generates and integrates emotional representations. Hence, illustrious cases of artistic imagination support the claim that creativity emanates in part from cognitive operations of representation combination.

Even kinesthetic creativity, the generation of novel forms of movement, can be understand as combination of representations as long as the latter are understood very broadly to include neural encodings of motor sequences (e.g., Wolpert & Ghahramani, 2000). Historically, novel motor sequences include the slam dunk in basketball, the over-the-shoulder catch in baseball, the Statue of Liberty play in football, the bicycle kick in soccer, and the pas de deux in ballet. All of these can be described verbally and may have been generated using verbal concepts, but it is just as likely that they were conceived and executed using motor representations that can naturally be encoded in patterns of neural activity.

We are primarily interested in creativity as a mental process, but we cannot neglect the fact that it also often involves interaction with the world. Manet's innovative painting arose in part from his physical interaction with the brush, paint, and canvas. Similarly, the invention of important technologies such as the stethoscope and the wheel can involve physical interactions with the world, as when Laennec rolled up a piece of paper to produce a hearing tube. External representations such as diagrams and equations on paper can also be useful in creative activities, as long as they interface with the internal mental representations that enable people to interact with the world.

We have provided examples to show that combination of representations is a crucial part of creativity in the domains of scientific discovery, technological invention, social innovation, and artistic imagination. Often these domains intersect, for example, when the discovery of electromagnetism made possible the invention of the radio, and when the invention of the microscope made possible the discovery of the cell. Social innovations can involve both scientific discoveries and technological invention, as when public health is fostered by the germ theory of disease and the invention of antibiotics. Artistic imagination can be aided by technological advances, for example, the invention of new musical instruments.

We hope that our examples make plausible the hypothesis that creativity often requires the combination of mental representations operating with multiple modalities. Chapter 9 provides much more extensive evidence that creativity in scientific discovery and technological invention results from combination of representations. We now describe neural mechanisms for such combinations.

Neural Combination and Binding

Combination of representations has usually been modeled with symbolic techniques common in the field of artificial intelligence. For example, concepts can be modeled by schemalike data structures called "frames" (Minsky, 1975), and combination of concepts can be performed by amalgamating frames (Thagard, 1988). Other computational models of conceptual combination have aimed at modeling the results of psycholinguistic experiments rather than creativity, but they also take concepts to be symbolic, verbal structures (Costello & Keane, 2000). Rule combination has been modeled with simple rules consisting of strings of bits and genetic algorithms (Holland et al., 1986), and genetic algorithms have also been used to produce new combinations of expressions written in the programming language LISP (Koza, 1992). Lenat and Brown (1984) produced discovery programs that generated new LISP-defined concepts out of chunks of LISP code. Rule-based systems such as ACT and SOAR can also employ learning mechanisms in which rules are chunked or compiled together to form new rules (Anderson, 1993; Laird, Rosenbloom & Newell, 1986).

These symbolic models of combination are powerful, but they lack the generality to handle the full range of representational combinations that include sensory and emotional information. Hence, we propose viewing representation combination at the neural level, since all kinds of mental representations—concepts, rules, sensory encodings, and emotions—are produced in the brain by the activity of neurons. Evidence for this claim comes from the vast range of psychological phenomena such as perception and memory that are increasingly being explained by cognitive neuroscience (see, e.g., Smith & Kosslyn, 2007; Chandrasekharan, 2009; Thagard, 2010a).

The basic idea that neural representations are constituted by patterns of activity in populations of neurons dates back at least to Donald Hebb (1949, 1980), and is implicit in many more recent and detailed neurocomputational accounts (e.g., Rumelhart & McClelland, 1986; Churchland, 1989; Churchland & Sejnowski, 1992; Dayan & Abbott, 2001; Eliasmith & Anderson, 2003). If this basic idea is right, then combination of representations should be a neural process involving generation of new patterns of activity from old ones.

Hebb is largely remembered today for the eponymous idea of learning of synaptic connections between neurons that are simultaneously active, which has its roots in the learning theories of eighteenth-century empiricist philosophers such as David Hartley. But Hebb's most seminal contribution was the doctrine that all thinking results from the activity of *cell assemblies*, which are groups of neurons organized by their synaptic connections and capable of generating complex behaviors.

Much later, Hebb (1980) sketched an account of creativity as a normal feature of cognitive activity resulting from the firing of neurons in cell assemblies. Hebb described problem solving as involving many "cell-assembly groups which fire and subside, fire and subside, fire and subside, till the crucial combination occurs" (Hebb, 1980, p. 119). Combination produces a new scientific idea that sets off a new sequence of ideas and constitutes a different way of seeing the problem situation by reorienting the whole pattern of cortical activity. The new combination of ideas that result from the connection of cell-assemblies forms a functional system that excites the arousal system, producing the *Eureka!* emotional effect. Thus Hebb sketched a neural explanation of the phenomenon of insight that has been much discussed by psychologists interested in problem solving (e.g., Bowden & Jung-Beeman, 2003; Sternberg & Davidson, 1995).

Hebb's conception of creative insight arising from cell assembly activity is suggestive but rather vague, and it raises as many questions as it answers. How are cell assemblies related to each other, and how is the information they carry combined? For example, if there is a group of cell assemblies (a neural population) that encodes the concept *sound*, and another that encodes the concept *wave*, how does the combined activity of the overall neural population encode the novel conceptual combination of *sound wave*?

We view the problem of creative combination of representations as an instance of the ubiquitous "binding problem" that pervades cognitive neuroscience (e.g., Treisman, 1996; Roskies, 1999). This problem was first recognized in studies of perception, where it is problematic how the brain manages to integrate various features of an object into a unified representation. For example, when people see a stop sign, they see the color red, the octagonal shape, and the white letters as all part of the same image, which requires the brain to bind together what otherwise might be several disparate representations. The binding problem is also integral to explaining the nature of consciousness, which has a kind of unity that may seem mysterious from the perspective of the variegated activity of billions of neurons processing many different kinds of information.

The most prominent suggestions of how to deal with the binding problem have concerned synchronization of neural activity, which has been proposed as a way to deal with the kind of cognitive coordination that occurs in consciousness (Crick, 1994; Engel, et al. 1999; Grandjean, Sander & Scherer, 2008; Werning & Maye, 2007). At a more local level, neural synchrony has been proposed as a way of integrating crucial syntactic information needed for the representation of relations, for example, to mark the difference between *Romeo loves Juliet* and *Juliet loves Romeo* (Shastri & Ajjanagadde, 1993; Hummel & Holyoak, 1997, 2003). For the first sentence, the neural populations for *Romeo* and *SUBJECT* would be active for a short period of time, then the neural populations for *loves* and *VERB*, and finally for *Juliet* and *OBJECT*. For the second sentence, the timing would be changed so that *Romeo* and *OBJECT* were synchronized, as well as *Juliet* and *SUBJECT*.

Unfortunately, there has been little success at finding neural mechanisms that could cause this synchronization behavior. Simple approaches, such as having a separate group of neurons that could specifically stimulate pairs of concepts, do not scale up, as they must have neurons for every single possible conceptual combination and thus require more neurons than exist in the human brain. Other models restrict themselves to particular modalities, such as binding color and location (Johnson, Spencer & Schoner, 2008). Still others are too brittle, assuming that neurons have no randomness and never die (for a detailed discussion, see Stewart & Eliasmith, 2009a; Eliasmith & Stewart, forthcoming).

Our aim in this chapter is to develop an alternative account based on convolution rather than synchronization. To do this, we present a neurally realistic model of a mechanism whereby arbitrary concepts can be combined, resulting in a new representation. We will not propose a general theory of binding, but rather will defend a more narrow account of how the information encoded in patterns of neural activity gets combined into new representations that may turn out to be creative. We draw heavily on Eliasmith's (2004, 2005a, forthcoming) work on neurobiologically plausible simulations of high-level inference.

Binding by Convolution

Tony Plate (2003) developed a powerful way of thinking about binding in neural networks, using vector-based representations similar to but more computationally efficient than the tensor-product representations proposed by Smolensky (1990). Our presentation of Plate's idea, which we call "binding by convolution," will be largely metaphorical in this section, but technical details are provided later. Eliasmith and Thagard (2001) provide a relatively gentle introduction to Plate's method.

To get the metaphors rolling, consider the process of braiding hair. Thousands of long strands of hair can be braided together by twisting them systematically into one or more braids. Similar twisting can be used to produce ropes and cables. Another word for twisting and coiling up things in this way is "convolve," and things are convolved if they are all twisted up together. Another word for "convolve" is "convolute," but we rarely use this term because in recent decades the term "convoluted" has come to mean "excessively complicated."

In mathematics, a convolution is an integral function that expresses the amount of overlap of one function f as it is shifted over another function g, expressing the blending of one function with another (see http://mathworld.wolfram.com/Convolution.html). This notion gives a mathematically precise counterpart to the physical process of braiding, as we can think of mathematical convolution as blending two signals together (each represented by a function) in a way roughly analogous to how braiding blends strands of hair together.

Plate developed a technique he called "holographic reduced representations" that applies an analogue of convolution to vectors of real numbers. It is natural to think of patterns of neural activity using vectors: If a neural population contains n neurons, then its activity can be represented by a sequence that contains n numbers, each of which stands for the firing rate of a neuron. For example, if the maximum firing rate of a neuron is 200 times per second, the rate of a neuron firing 100 times per second could be represented by the number 0.5. Then the vector (0.5, 0.4, 0.3, 0.2, 0.1) corresponds to the firing rates of this neuron and four additional ones with slower firing rates.

So here is the basic idea: If we abstractly represent the pattern of activity of two neural populations by vectors A and B, then we can represent their combination by the mathematical convolution of A and B, which is another vector corresponding to a third pattern of neural activity. For the moment, we ignore what this amounts to physiologically—see the simulation section below. The resulting vector has emergent properties, that is, properties not possessed by (or simple aggregates of) either of the two vectors out of which it is combined. The convolved vector combines the information included in each of the originating vectors in a nonlinear fashion that enables only approximate reconstruction of them. Hence the convolution of vectors produces an emergent binding, one that is not simply the sum of the parts bound together (on emergence, see Bunge, 2003, and Wimsatt, 2007).

Talking about convolution of vectors still does not enable us to grasp the convolution of patterns of neural activity. To explain that, we will need to describe our simulation model of how combination can occur in a biologically realistic, computational neural network. Before getting into technical details, however, we need to describe how the *Aha!* experience can be understood as convolution of a novel combined representation with patterns of brain activity for emotion.

Emotion and Creativity

Cognitive science must explain not only how representations get combined into creative new ones, but also how such combinations can be intensely emotional. Many quotes from eminent scientists attest to the emotional component of scientific discovery, including such reactions as delight, amazement, pleasure, glory, passion, and joy (Thagard, 2006a, ch. 10). We expect that breakthroughs in technological invention, social innovation, and artistic imagination are just as exciting. For example, Richard Feynman (1999, p. 12) wrote: "The prize is the pleasure of finding a thing out, the kick of the discovery, the observation that other people use it [my work]—those are the real things, the others are unreal to me." What are the neuropsychological causes of the "kick of the discovery"?

To answer that question, we need a neurocomputational theory of emotion that can be integrated with the account of representation generation provided earlier in this chapter. Thagard and Aubie (2008) have hypothesized how emotional experience can arise from a complex neural process that integrates cognitive appraisal of a situation with perception of internal physiological states. Figure 8.1 shows the structure of the EMOCON model, with the interaction of multiple brain areas generating emotional consciousness, requiring both appraisal of the relevance of a situation to an agent's goals (largely realized by the prefrontal cortex and mid-brain dopamine system) and internal perception of physiological changes (largely realized by the amygdala and insula). For a defense of the neural, psychological, and philosophical plausibility of this account of emotional experience, see also Thagard (2010a).

If the EMOCON model is correct, then emotional experiences such as the ecstasy of discovery are patterns of neural activity, just like other mental representations such as concepts and rules. Now it is easy to see how the *Aha!* or *Eureka!* experience can arise. When two representations are combined by convolution into a new one, the brain automatically performs an evaluation of the relevance of the new representation to its goals. Ordinarily, such combinations are of little significance, as in the ephemeral conceptual combinations that take place in all language processing. There need be no emotional reaction to mundane combinations such as *brown cow* and *tall basketball player*. But some combinations are surprising, such as *cow basketball*, and may elicit further processing to try to make sense of them (Kunda, Miller & Claire, 1990).

In extraordinary situations, the novel combination may be not only surprising but actually exciting, if it has strong relevance to accomplishing the longstanding goals of the thinker. For example, Darwin was thrilled when he realized that the novel combination *natural selection* could explain facts about species that had long puzzled him, and the combination *wireless email* excited the inventors of the Blackberry when they realized its great commercial potential.

Figure 8.2 shows how representation combination can be intensely emotional, when patterns of neural activity corresponding to concepts become convolved with patterns of activity that constitute the emotional evaluation of the new combination. A new combination such as *sound wave* is exciting because it is highly relevant to accomplishing the discoverer's goals. But emotions are not just a purely cognitive process of appraisal, which could be performed dispassionately, as in a calculation of expected utility as performed by economists. *Aha!* is a very different experience from "Given the probabilities and expected payoffs, the expected value of option X is high."

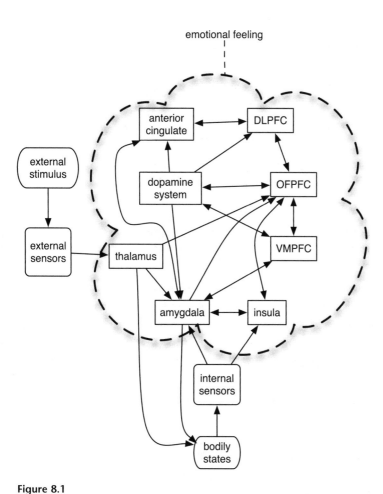

Figure 8.1
The EMOCON model of Thagard and Aubie (2008), which contains details and partial computational modeling. Abbreviations: DLPFC is dorsolateral prefrontal cortex. OFPFC is orbitofrontal prefrontal cortex. VMPFC is ventromedial prefrontal cortex. The dotted line is intended to indicate that emotional consciousness emerges from activity in the whole system.

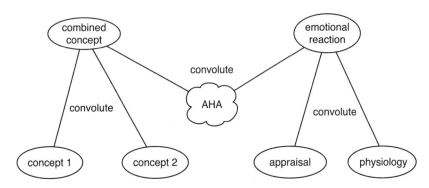

Figure 8.2
How the *Aha!* experience arises by multiple convolutions of representation combination and emotion. Figure 8.9 presents a neural version.

Physiology is a key part of emotional experience—racing heartbeats, sweaty palms, and so on, as pointed out by many theorists (e.g., James, 1884; Damasio, 1994; Prinz, 2004). But physiology cannot be the only part of the story, as there are many reasons to see cognitive appraisal as crucial too (Thagard, 2010a). The EMOCON model shows how reactions can combine both cognitive appraisal and physiological perception. Hence we propose that the *Aha!* experience requires a triple convolution, binding: (1) two representations into an original one, (2) cognitive appraisal and physiological perception into a combined assessment of significance, and (3) the combined representation and the integrated cognitive/physiological emotional response into a unified representation (pattern of neural activity) of the creative representation and its emotional value.

Because the brain's operations are highly parallel, it would be a mistake to think of what we have just described as a serial process of first combining the representations, then integrating the appraisal and physiological perception, and finally convoluting the new representation and the emotional reaction. Rather, all these processes take place concurrently, with a constant flow of activation among the regions of the brain crucial for different kinds of cognition, perception, and emotion. The *Aha!* experience seems mysterious because we have no conscious access to any of these processes or their integration. But figure 8.2 shows a possible mechanism for how the wonderful *Aha!* experience can emerge from neural activity.

Simulations

Our discussion of convolution so far has been largely metaphorical. We will now describe neurocomputational simulations that use the methods of Eliasmith (2004, 2005a) to show: (1) how patterns of neural activity can represent vectors; (2) how convolution can bind together two representations to form a new one (and subsequently unbind them); and (3) how convolution can bind together a new combined representation with patterns of brain activity that correspond to emotional experience arising from a combination of cognitive appraisal and physiological perception.

Simulation 1: Neurocomputational Model of Visual Patterns

The first requirement for a mechanistic neural explanation of conceptual combination is specifying how a pattern can be represented by a population of neurons. Earlier, we gave a simple example where a vector of five elements corresponded to the firing rates of five neurons; for example, the value 0.5 could be indicated by an average firing rate of 100Hz (i.e., 100 times per second). Linguistic representations can be translated into vectors using Plate's method of holographic reduced representation, and there are also natural translations of visual, auditory, and olfactory information into the mathematical form of vectors. So, a general multimodal theory of representation combination need consider only how vectors can be neurally represented and combined.

We can depict the simple approach visually as in figure 8.3. Here, we are using 25 neurons (the circles) to represent a 25-dimensional vector (squares). The shading of each neuron corresponds to its firing rate (with white being fast and black being slow). Depending on the firing rate, these same neurons can represent different vectors, and three different possibilities are shown.

Although this approach for representation is simple, it is not biologically realistic. In brains, if we map a single value onto a single neuron, then the randomness of that neuron will limit how accurately that value can be represented. Neurons are highly stochastic devices, and can even die, so we cannot use a method of representation that does not have some form of redundancy. To deal with this problem, we need to use many more neurons to represent the same vector. In figure 8.4, we

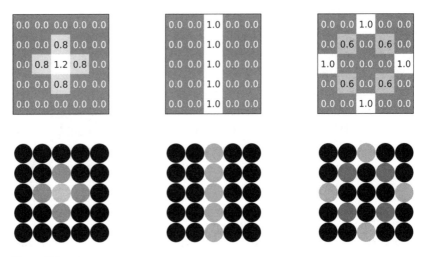

Figure 8.3
Simple neural network encoding using one neuron per pixel value. The three numerical vectors are shown at the top, shaded by value to aid interpretation. At the bottom, each circle is an individual neuron and the lighter shading indicates a faster firing rate. Encodings for three different 25-dimensional vectors are shown.

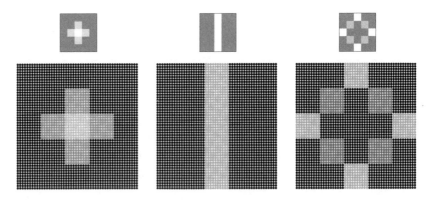

Figure 8.4
Simple neural network encoding using 100 neurons per pixel value. The three vectors are shown at the top (see figure 8.3 for more details). In the large bottom squares, each circle is an individual neuron and the lighter shading indicates a faster firing rate. Encodings for three different 25-dimensional vectors are shown.

represent the same three patterns as figure 8.3, but using 100 times as many neurons.

This approach allows for increased accuracy and robustness to neuron death, but it again does not correspond to what is found in real brains. Instead of having a large number of neurons, each of which behaves similarly to its neighbors (as do each of the groups of 100 neurons in figure 8.4), there is strong neurological evidence that vectors are represented by each neuron having a different overall pattern in response to which they fire most quickly. For example, Georgopoulos, Schwartz, and Kettner (1986) demonstrated that neurons in the motor cortex of monkeys encode reaching direction by each one having its own preferred direction vector. That is, for each neuron, there is a particular vector for which it fires most quickly, and the firing rate decreases as the vector represented becomes more dissimilar to that direction vector. As another example, neurons in the visual cortex respond to patterns over the visual field, and each one responds most strongly to a slightly different pattern, with neurons that are near each other responding to similar patterns (e.g., Blasdel & Salama, 1986).

To adopt this approach, instead of having 100 neurons responding to each single value in the vector, we take each neuron and randomly choose a particular pattern for which it will fire the fastest. For other vectors, it will fire less quickly, based on the similarity with its preferred vector. This approach corresponds to that seen in many areas of visual and motor cortex, and has been shown to allow for improved computational power over the previous approach. Figure 8.5 shows a group of 2,500 neurons representing three 25-dimensional vectors in this manner. It should be noted that, although we can no longer visually see the relationship between the firing pattern and the original vector (as we could in figure 8.4), the vector is still being represented by this neural firing, as we can take this firing pattern and derive the original pattern, if we know the preferred direction vectors for each of these neurons (for more details, see the mathematical appendix to Thagard & Stewart, 2011).

To further improve the biological realism of our model, we can construct our model using spiking neurons. That is, instead of just calculating a firing rate, we can model the flow of current into and out of each neuron, and when the voltage reaches its firing potential (around –45mV), it fires. Once a neuron fires, it returns to its resting potential (–70mV) and excites all the

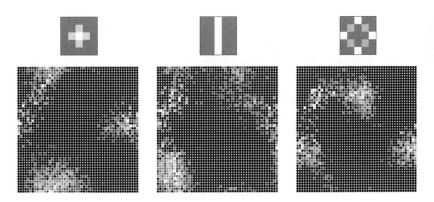

Figure 8.5
Distributed encoding using preferred direction vectors of a 25-dimensional vector
with 2,500 neurons. Each circle is an individual neuron and the lighter shading
indicates a faster firing rate. Encodings for three different 25-dimensional vectors
are shown.

neurons to which it is connected, thus changing the current flows in those
neurons. With these neurons, we set the amount of current flowing into
them (based on the similarity between the input vector and the preferred
direction vector), instead of directly setting the firing rate.

The simplest model of this process is the Leaky Integrate-and-Fire
neuron. We use it here, although all of the techniques discussed in this
chapter can be applied to more complex neural models. If we run the
model for a very long time and average the resulting firing rate, we get
the same picture as in figure 8.5. However, at any given time within the
simulation, the voltage levels of the various neurons will be changing.
Figure 8.6 shows a snapshot of the resulting neural behavior, where black
indicates the neuron resting and white indicates that enough voltage has
built up for the neuron to fire. Any neuron that is white in this figure will
soon fire, return to black, and then slowly build up voltage over simulated
time. This rate is governed by the neuron's membrane time constant (set
to 20ms in accordance with McCormick et al., 1985) and the refractory
period (2ms).

Simulation 2: Convolution of Patterns
Now that we have specified how neurons can represent vectors, we can
organize neurons to perform the convolution operation (Eliasmith, 2004).

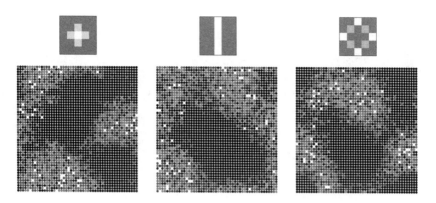

Figure 8.6
A snapshot in time showing the voltage levels of 2,500 neurons using distributed encoding of preferred direction vectors of a 25-dimensional vector. Each circle is an individual neuron and the lighter shading indicates a higher membrane voltage. White circles indicate a neuron that is in the process of firing. Encodings for three different 25-dimensional vectors are shown.

We need a neural model where we have two groups of neurons representing the input patterns (using the representation scheme above), and these neurons must be connected to a third group of neurons which will be driven to represent the convolution of the two original patterns.

The Neural Engineering Framework (NEF) of Eliasmith and Anderson (2003) provides a methodology for converting a function such as convolution into a neural model by deriving the synaptic connection weights that will implement that function. Once these weights are found, we use the representation method discussed above to encode the two original vectors in neural groups A and B. When these neurons fire, the synaptic connections cause electric current to flow into any neurons to which they are connected. This flow in turn causes firing in neural group C, which will represent the convolution of the patterns in A and B. (Details on the derivation of these synaptic connections can be found in the appendix to Thagard & Stewart 2011.) Figure 8.7 shows how convolution combines the neural representation of two perceptual inputs, on the left, into a neural representation of their convolution, on the right. It is clear from the visual interpretation of the neural representation on the right that the convolution of the two input patterns is not simply the sum of those patterns, and therefore amounts to an emergent binding of them.

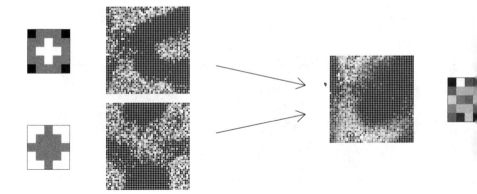

Figure 8.7
Convolution occurring in simulated neurons. The two input grids on the left are represented by the neural network firing patterns beside them. On the right is the neural network firing pattern that represents the convolution of the two inputs. The grid on the far right is the visual interpretation of the result of this convolution. Arrows indicate synaptic connections via intervening neural populations.

Importantly, one set of neural connection weights is sufficient for performing the convolution of *any* two input vectors. That is, the synaptic connections do not need to be changed if we need to convolve two new patterns; all that has to change is the firing patterns of the neurons in groups A and B. We do not rely on a slow, learning process of changing synaptic weights: convolution is a fast process response to changes in perceptual inputs. If the synaptic connections in our NEF model correspond to a fast neurotransmitter, convolution can occur within five milliseconds. However, we make no claims as to *how* the neurons come to have the particular connection weights that allow them to perform this convolution. These weights could be genetically specified or could be learned over time.

After two representations have been combined, it is still possible to extract the original information from the combined representation. The process of convolution can be reversed, using neural connections almost identical to those needed for performing the convolution in the first place, as shown in figure 8.8. There is a loss of information in that the extracted information is only an approximation of the original. However, by increasing the number of vector values and the number of neurons per value, we

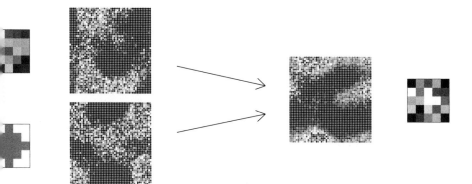

Figure 8.8
Deconvolution occurring in simulated neurons. Inputs are (top left) the output from
figure 8.7 and (bottom left) the other input from figure 8.7. The result (far right) is
an approximation of the first of the original inputs (top left in figure 8.7).

can make this approximation as accurate as desired (Eliasmith & Anderson,
2003). Importantly, this is not a selective process: all of the original patterns are preserved. Given the concept *sound wave*, we can always break it
back down into *sound* and *wave*. Any specialization of the new concept
must involve the development of new associations with the new pattern.
The process for this is outside the scope of this chapter, but since the new
pattern is highly dissimilar from the original patterns, these new associations can be formed without disrupting conceptual associations with the
original patterns.

The key point here is that the process of convolution generates a new
pattern given any two previous patterns, and that this process is reversible.
In figure 8.7, the pattern on the right bears no similarity to either of the
patterns on the left. In mathematical terms, the degree of similarity can
be measured using the dot product, which tends toward zero the larger the
vectors (Plate, 2003). This means that the new pattern can be used for
cognitive processing without being mistaken for existing patterns.

However, this new pattern can be broken back down into an approximation of the original patterns, if needed. This allows us to make the claim
that the representational content is preserved via convolution, at least to
a certain degree. The pattern on the right of figure 8.8 bears a close resemblance to the pattern on the top left of figure 8.7, and this accuracy

increases with the number of dimensions in the vector and the number of neurons used. Stewart, Choo, and Eliasmith (2010b) have created a large-scale (373,000-neuron) model of the basal ganglia, thalamus, and cortex using this approach, and have shown that it is capable of taking the combined representation of a sentence (such as "Romeo loves Juliet") and accurately answering questions about that sentence. This result demonstrates that all of the representational content can be preserved up to some degree of accuracy. Furthermore, they have also shown that these representations can be manipulated using the equivalent of if-then production rules that indicate which concepts should be combined, and which ones should be taken apart. When these rules are implemented using realistic spiking neurons, they are found to require approximately 40 milliseconds for simple rules (not involving conceptual combination) and approximately 65 milliseconds for complex rules that combine concepts or extract them (Stewart, Choo & Eliasmith, 2010a). This finding accords with the standard empirical finding that humans require 50 milliseconds to perform a single cognitive action. However, this research has not addressed how different modalities can be combined.

Simulation 3: Multimodal Convolution

Simulation 2 demonstrates the use of biologically realistic artificial neurons to combine two arbitrary vector-based representations, showing the feasibility of convolution as a method of conceptual combination. We used a visual example, but many other modalities can be captured by vectors. Hence, we can create multimodal representations by convolving representations from distinct modalities. For example, we might combine a particular visual stimulus with a particular auditory stimulus, along with a symbolic label, and even an emotional valence.

Earlier we proposed that the *Aha!* experience required a convolution of at least four components: two representations that get combined into a new one, and two aspects of emotional processing—cognitive appraisal and physiological perception. Figure 8.9 shows our NEF simulation of how this might work. Instead of just two representations being convolved, there are a total of four that could stand for various aspects of the cognitive and emotional content of a situation. Each convolution is implemented as before, and they are added by having all convolutions project to the same set of output neurons.

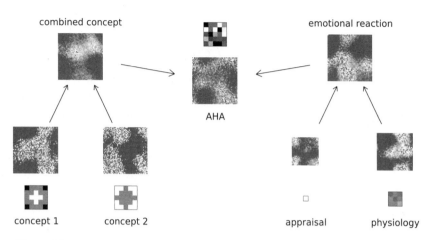

combined concept   emotional reaction

AHA

concept 1   concept 2   appraisal   physiology

Figure 8.9
Combining four representations into a single result. The arrows suggest the main
flow of information, but do not preclude the presence of many reentry (feedback)
loops.

We do not need to assume that all of the vectors being combined are
of the same length. For example, representing physiological perception
may only require a few dimensions, whereas encoding symbolic content
requires hundreds of dimensions. These vectors can still be combined by
projecting the low-dimensional vector into the higher-dimensional space.
This means that we can combine any representations of any size and any
modality into a single novel representation.

In our simulations, we have shown that vectors can be encoded using
neural firing patterns. These vectors can represent information in any
modality, from a visual stimulus to an internal state to a symbolic term.
Furthermore, these representations can be combined via a neural implementation of the convolution operation that produces a new vector that
encodes an approximation of all the original vectors. We have thus shown
the computational feasibility of using convolution to combine representations of concepts and emotional reactions to them, generating the *Aha!*
experience.

Our examples of convolution using visual patterns are obviously much
simpler than important historical cases such as the double helix and the
BlackBerry. But complex representations are made up of simpler ones, and
convolution provides a plausible mechanism for building the kinds of rich,
multimodal structures needed for creative human cognition.

What Convolutions Are Creative?

Obviously, however, not all convolutions are creative. If our neural account is correct, even mundane conceptual combinations such as *red shirt* are produced by convolution, so what makes some convolutions creative? According to Boden (2004), what characterizes acts as creative is that they are novel, surprising, and valuable. Most combinations of representations meet none of these conditions. However, convolutions that generate the *Aha!* emotional response are far more likely to satisfy them, as can be shown by consideration of the relevant affective mechanisms.

In our current simulations, the appraisal and physiological components of emotion are provided as inputs, but they could naturally be generated by the neural processes described in the EMOCON model of Thagard and Aubie (2008). Following Sander, Grandjean, and Scherer (2005), novelty could be assessed by inputs concerning suddenness, familiarity, predictability, intrinsic pleasantness, and goal-need relevance. The latter two factors, especially goal-need relevance, are also directly relevant to assessing whether a new conceptual combination (new convolution) is potentially valuable. A mundane combination such as *red shirt* generates little emotion because it usually makes little contribution to potential goal satisfaction. In contrast, combinations like *wireless email* and *sound wave* can be quickly appraised as highly relevant to satisfying commercial or scientific goals. Hence such convolutions get stronger emotional associations because the appraisal process has marked them as novel and valuable, in line with Boden's criteria.

Surprise is more complicated. Thagard (2000) proposed a localist neural network model of surprise in which an assessment is made of the extent to which nodes change their activation. Surprise concerning an element results from rapid shifts in activation of the node representing the element from positive to negative or vice versa. A similar process could be built into a more neurologically plausible distributed model by having neural populations that respond to rapid changes in the patterns of activation in other neural populations. Mathematically, this process is equivalent to taking the derivative of the activation value, and realistic neural models of this form have been developed (Tripp & Eliasmith, 2010).

Like other emotional reactions, surprise comes in various degrees, from mild appreciation of something new to wild astonishment. Thagard and

Aubie (2008) propose that the intensity of emotional experience derives from the firing rates of neurons in the relevant populations. For example, high degrees of pleasure arise from rapid rates of firing in neurons in the dopamine system. Analogously, a very strong *Aha!* reaction would result from rapid firing in the neural population that provides the convolution of the conceptual combination with the emotional reaction that itself combines both cognitive appraisal and physiological perception. There is no need to postulate some kind of *Aha!* detection module in the brain, as the emotion-generating convolutions may occur on the fly in various brain areas with connectivity to neural populations for both representing and evaluating. The representational format for all these processes is the same—patterns of neural firing—so the exchange and integration of many different kinds of information is achieved.

In these ways, the neural system itself can emotionally mark some convolutions as representing results that are potentially more novel, surprising, and valuable and hence more likely to qualify as creative according to Boden's (2004) criteria. Of course, there are no guarantees, as people sometimes get excited about ideas that turn out to be derivative or useless. But it is a crucial part of the emotional system that it serves to identify some conceptual combinations as exciting and hence worth storing in long-term memory and serving in future problem solving. The *Aha!* experience is not just a side effect of creative thinking, but rather a central aspect of identifying those convolutions that are potentially creative.

A philosophical worry about combination and convolution is how newly generated patterns of neural activation can retain or incorporate the content (meaning) of the original concepts. How does the convolved representation for *wireless email* contain the content for both *wireless* and *email*? Answering this question presupposes solutions to highly controversial philosophical problems about how psychological and neural representations gain their contents. We cannot defend a full answer here, but merely provide a sketch based on accounts of neurosemantics defended elsewhere (Eliasmith, 2005b; Parisien & Thagard, 2008; Thagard, 2010a). We avoid the term "content" because it misleadingly suggests that the meaning of a representation is some kind of thing rather than a multifaceted relational process.

Representations acquire meaning in two ways, through processes that relate them to the world and through processes that relate them to other representations. A representational neural population is most clearly meaningful when its firing activity is causally correlated with events in the world, that is, when there is a statistical correlation that results from causal interactions. These interactions can operate in both directions, from perception of objects in the world causing neural firing, and from neural firing causing changes in perceptions of objects. However, the firing activity of a representational neural population is also affected by the firing activity of other neural populations, as is most clear in neural populations for abstract concepts such as *justice* and *infinity*. Hence the crucial question about the meaningfulness of conceptual combinations by convolution is: How do the relational processes that establish the meaning of two original concepts contribute to the relational processes that establish the meaning of the combined concept produced by convolution in a neural network?

Take a simple example such as *red shirt*. The neural representation (firing pattern) for *red* gets its meaning from causal correlations with red things in the world and with causal correlations with other neural representations such as ones for *color* and *blood*. Similarly, the neural representation for *shirt* gets its meaning from causal correlations with shirts and with other neural representations such as ones for *clothing* and *sleeves*. If the combination *red shirt* was triggered by perception of a red shirt, then the new convolving neural population will have a causal correlation with the stimulus, just as the neural populations for *red* and *shirt* will. Hence there will be some overlap in the firing patterns for *red*, *shirt*, and *red shirt*. Similarly, if the conceptual combination is triggered by an utterance of the words "red shirt" rather than anything perceptual, there will still be some overlap in the firing patterns of *red shirt* with *red* and *shirt* thanks to interconnections with the other related concepts such as *color* and *clothing*. Thus it is reasonable to conclude that the neural population representing the convolution *red shirt* retains much of the meaning of both *red* and *shirt*.

Limitations

We have presented a detailed, neurocomputational account of psychological mechanisms that may contribute to creativity and the *Aha!* experience. Our hypothesis that creativity arises from neural processes of convolution explains how multimodal concepts in the brain can be combined, how original ideas can be generated, and how production of new ideas can be an emotional experience.

But we acknowledge that our account has many limitations with respect to describing the neural and other kinds of processes that are involved in creativity and innovation. We see the models we have presented here as only a small part of a full theory, so we will now sketch what we see as some of the missing neural, psychological, and social ingredients.

First, we have no direct evidence that convolution is the specific neuro-computational mechanism for conceptual combination. However, it is currently the only proposed mechanism that is general purpose, scalable, and works with realistic neurons. Convolution has the appropriate mathematical properties for combining any sort of neural information and is consistent with biological limitations. It scales up to human-sized vocabularies (Plate, 2003) without requiring an unfeasibly large number of additional neurons to coordinate firing activity in different neural populations (Eliasmith & Stewart, forthcoming). The mechanism works with noisy spiking neurons, and can make use of detailed models of individual neurons. That said, we need additional neuroscientific evidence and to develop alternative computational models before it would be reasonable to claim that convolution is *the* mechanism of idea generation. Theorizing about neural mechanisms for high-level cognition is still in its infancy. We have assumed, in accord with many current views in theoretical neuroscience, that patterns of neural firing are the brain vehicles for mental representations, but future research may necessitate a more complex account that incorporates, for example, chemical activity in glial cells that interact with neurons.

Second, our current model of convolution combines whole concepts without selectively picking out aspects of them. In contrast, the symbolic computational model of conceptual combination developed by Thagard (1988) allows new concepts, construed as a collection of slots and values, to select various slots and values from the original concepts. We conjecture that convolution can be made more selective using mechanisms like those now performing neurobiologically realistic rule-based reasoning (Stewart, Choo & Eliasmith, 2010b), but selective convolution is a subject for future research.

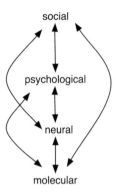

Figure 8.10
Causal interactions between four levels of analysis relevant to understanding creativity.

More generally, our account is intended as only part of a general, multi-level account of creativity. By no means are we proposing a purely neural, ruthlessly reductionist account of creativity. We recognize the importance of understanding thinking in terms of multilevel mechanisms, ranging from the molecular to the psychological to the social, as shown in figure 8.10 (Thagard, 2009, 2010a; see also Bechtel, 2008, and Craver, 2007). Creativity has many social aspects, requiring interaction among people with overlapping ideas and interest. For example, scientific research today is largely collaborative, and many discoveries occur because of fruitful interactions among researchers (Thagard, 1999, ch. 11; Thagard, 2006b). Neurocomputational models ignore the social causes of creative break-throughs, which are often crucial in explaining how different representations come to be combined in a single brain. For example, Darwin's ideas about evolution and breeding (artificial selection) were the result of many interactions he had with other scientists and farmers, and various engineers were involved in the development of the wireless technologies that evolved into the BlackBerry. A full theory of creativity and innovation will have to flesh out the upward and downward arrows in figure 8.10 in a way that produces a more complete account of creativity.

Figure 8.10 is incompatible with other, more common views of causality in multilevel systems. Reductionist views assume that causality runs only upward, from the molecular to the neural to the psychological to the social. At the opposite extreme, antireductionist views assume that the complexity of systems and the nature of emergent properties (ones that hold of higher-level objects and not of their constituents) means that higher levels must be described as largely independent of lower ones, so that social explanations (e.g., in sociology and economics) have a large degree of autonomy from psychological ones, and psychological explanations have a large degree of autonomy from neural and molecular ones.

From our perspective, reductionist, individualistic views are inadequate because they ignore ways in which objects and events at higher levels have causal effects on objects and events and lower levels. Consider two scientists (perhaps Crick and Watson, who developed many of their ideas about DNA jointly), managing in conversation to make a creative breakthrough. This conversation is clearly a social interaction, with two people communicating in ways that may be both verbal and nonverbal. Such interactions can have profound psychological, neural, and molecular effects. When the scientists bring together two concepts or other representations that they have not previously connected, their beliefs may change dramatically, for example, when they realize they have found a hypothesis that can solve the problem on which they have been jointly working. This psychological change is also a neural change, altering their patterns of neural activity. If the scientists are excited or even a little pleased by their new discovery, they will also undergo molecular changes such as increases in dopamine levels in the nucleus accumbens and other reward-related brain areas. Because social interactions can cause important psychological, neural, and molecular changes, the reductionist view that only considers how lower-level systems causally affect higher-level ones is implausible.

On the other hand, the holistic, antireductionist view that insists on autonomy of higher levels from lower ones is also implausible. Molecular changes even as simple as ingestion of caffeine or alcohol can have large effects on psychological and social processes; compare the remark that a mathematician is a device for turning coffee into theorems. If our account of creative representation combination as neural binding is on the right track, then part of the psychological explanation of the creativity of individuals, and hence part of the social explanation of the productivity of groups, will require attention to neural processes. Genetic processes may also be relevant, as seen in the recent finding that combinations of genes involved in dopamine transmission have some correlation with artistic capabilities (Kevin Dunbar, personal communication).

One useful way to characterize multilevel relations is provided by Bunge (2003), who critiques both holism and individualism. He defines a system as a quadruple:

<Composition, Environment, Structure, Mechanism> where:

Composition = collection of parts;

Environment = items that act on the parts;

Structure = relations among parts, especially bonds between them;

Mechanism = processes that make the system behave as it does.

Our discussion in this chapter has primarily been at the neural level: the composition is a collection of neurons, the environment consists of the physiological inputs such as external and internal perception that cause changes in firing of neurons, the structure consists of the excitatory and inhibitory synaptic connections among neurons, and the mechanism is the whole set of neurochemical processes involved in neural firing.

A complete theory of creativity would require specifying social, psychological, and molecular systems as well, not just on their own, but in relation to neural processes of creativity. We would need to specify the composition of social, psychological, neural, and molecular systems in a way that exhibits their part-whole relations. Much more problematically, we need to describe the relations among the processes that operate at different levels. This project of multilevel interaction goes far beyond the scope of the current chapter, but we mention it here to forestall the objection that neural convolution is only one aspect of creativity: We acknowledge that our neural account is only part of a full scientific explanation of creativity. Thagard (2010d, forthcoming-c) advocates the method of multilevel interactive mechanisms (MIMs) as a general approach for cognitive science.

Various writers on the social processes of innovation have remarked on the importance of interpersonal contact for the transmission of *tacit knowledge*, which can be very difficult to put into words (e.g., Asheim & Gertler, 2005). Our neural account provides a nonmysterious way of understanding tacit knowledge, as the neural representations our models employ are compatible with procedural, emotional, and perceptual representations that may be nonverbal. Transferring such information may require thinkers to work together in the same physical environment so that bodily interactions via manipulation of objects, diagrams, gestures, and facial expressions can provide sources of communication that may be as rich as verbal conversation. A full account of creativity and innovation that includes the social dimension should include an explanation of how nonverbal communication can contribute to the joint production of tacit knowledge, which we agree is an important part of scientific thinking (Sahdra & Thagard, 2003). See Hélie and Sun (2010) for a connectionist account of insight through conversion of knowledge from implicit to explicit.

Even at the psychological level, our account of creativity is incomplete. We have described how two representations can be combined into new ones, but we have not attempted to say what triggers such combinations. Specifying triggering conditions for representation combination would require full theories of language processing and problem solving. We have not investigated how conceptual combination can occur as part of language processing (e.g., Medin & Shoben, 1988; Wisniewski, 1997). Conceptual combination is clearly not sufficient for creativity, as people make many boring combinations as part of everyday communication. Moreover, in addition to conceptual combination, there are other kinds of mental operations important for creativity, as we review in the next section on comparisons with other researchers' accounts of creativity.

For problem solving, Thagard (1988) presented a computational model of how conceptual combination could be triggered when two concepts become attributed to the same object during an attempt to solve explanation problems, and a similar account could apply to the kinds of creativity we have been concerned with here. When scientists, inventors, social activists, or artists are engaged in challenging activity, they naturally have mental representations such as concepts active simultaneously in working memory. Combining such representations occasionally produces creative results. But the neurocomputational models described in this chapter deal only with the process of combination, not with triggering conditions for combination or with postcombination employment of newly created and combined representations.

A full account of creativity as representation combination would need to include both the upstream processes that trigger combination and the downstream processes that make use of newly created representations, as shown in figure 8.10. The downstream processes include assessing the ongoing usefulness of the newly created representations for whatever purpose inspired them. For example, the creation of a new theoretical concept such as *sound wave* becomes important if it enters the scientific

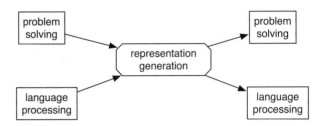

Figure 8.11
Problem solving and language processing as contributing to and affected by representation combination.

vocabulary and becomes subsequently employed in ongoing explanations. Thus a full theory of creativity would have to include a description of problem solving and language processing as both inputs to and outputs from the neural process of representation generation, which is all we have tried to model in this chapter. Implementing figure 8.11 would require integrated neurocomputational models of problem solving and language processing that remain to be developed.

Comparisons with Related Work

Our neural models of representation combination and emotional reaction are broadly compatible with the ideas of many researchers on creativity whose work has addressed different issues. Boden (2004) usefully distinguishes three forms of creativity: combinatorial, exploratory, and transformational. We have approached only the combinatorial form in which new concepts are generated, but have not dealt with the broader exploration and transformation of conceptual spaces. We have been concerned with what Boden calls psychological creativity, which involves the generation of representations new to particular individuals, and not historical creativity, which involves the generation of completely new representations not previously produced by any individual.

Our approach in this paper has also been narrower than Nersessian's (2008) discussion of mental models and their use in creating scientific concepts. She defines a mental model as a "structural, behavioral, or functional analog representation of a real-world or imaginary situation, event, or process" (p. 93). We agree with her contention that many kinds of internal and external representations are used during scientific reasoning, including ones tightly coupled to embodied perceptions and actions. As yet, no full neurocomputational theory of mental models has been developed, so we are unable to situate our account of representation combination within the rich mental-models approach to reasoning. But our account is certainly compatible with Nersessian's claim (2008, p. 138) that conceptual changes arise from interactive processes of constructing, manipulating, evaluating, and revising models. Chapter 4 sketched how mental models can be understood in terms of neural processes.

We agree with Hofstadter (1995) that many computational models of analogy have relied too heavily on fixed verbal representations that prevent the kind of fluidity found in many kinds of creative thinking. In the models described in this chapter, we use neural representations of a more biologically realistic sort, representing concepts by the activity of thousands of spiking neurons, not by single nodes as in localist connectionist simulations, nor by a few dozen nonspiking neurons as in PDP simulations. We acknowledge, however, that we have not produced a new model of analogical processing. Eliasmith and Thagard (2001) made a first step toward a vector-based system of analogical mapping, but no full neural implementation of that system has yet been produced.

The scope of our project is also narrower than much artificial intelligence research on scientific discovery that develops more general accounts of problem solving (e.g., Langley et al., 1987; Bridewell et al., 2008). Our neural simulations are incapable of generating new scientific laws or representations of processes that are crucial for a general account of scientific discovery. Another example of representation generation is the "crossover" operation that operates as part of genetic algorithms in the generation of new rules in John Holland's classifier systems (Holland et al., 1986). Our models could be seen as doing a kind of crossover mating between neural patterns, but they are not embedded in a general system capable of making complex inferences.

Our account of creativity as based on representation combination is similar to the idea of blending (conceptual integration) developed by Fauconnier and Turner (2002), which is modeled computationally by Pereira (2007). Our account differs in providing a neural mechanism for combining multimodal representations, including emotional reactions.

Brain imaging is beginning to yield interesting results about the neural correlates of creativity (e.g., Bowden & Jung-Beeman, 2003; Kounios &

Beeman, 2009; Subranamiam et al., 2009). Unfortunately, our neural models of creativity are not yet organized into specific brain areas, so we cannot explain particular findings concerning these neural correlates.

Conclusion

Despite the limitations described in the last two sections, we think our account of representation generation is novel and interesting, perhaps even creative, in several respects. First, we have shown how conceptual combination can occur in biologically realistic populations of thousands of spiking neurons. Second, we employed a mechanism of binding—convolution—that differs in important ways from the synchrony mechanism that is more commonly advocated. Third, and perhaps most important, we have used convolution to show how the creative generation of new representations can generate emotional reactions such as the much-desired *Aha!* experience.

Convolution also provides an alternative to synchronization as a potential naturalistic solution to the classic philosophical problem of explaining the apparent unity of consciousness. In figure 8.8, we portrayed a simulation that integrates the activity of seven neural populations, showing how there can be a unified experience of the combination of two concepts and an emotional reaction to them. We do not mean to suggest that there is a specific locus of consciousness in the brain, as there are many different convergent zones (also known as association areas) where information from multiple neural populations can come together. The dorsolateral prefrontal cortex seems to be an important convergence zone for working memory and hence for consciousness, as in the EMOCON model of Thagard and Aubie (2008).

Neural processes of the sort we have described are capable of encoding the full range of representations that contribute to human thinking, including ones that are verbal, visual, auditory, olfactory, tactile, kinesthetic, procedural, and emotional. This range should enable application of the mechanism of representation combination to all realms of human creativity, including scientific discovery, technological invention, social innovation, and aesthetic imagination. But much research remains to be done to build a full, detailed account of the creative brain.

9 Creative Combination of Representations: Scientific Discovery and Technological Invention

Introduction

Human creativity operates in many domains, including scientific discovery, technological invention, artistic imagination, and social innovation. What are the cognitive processes that produce these creative results? Are there psychological mechanisms common to such diverse products of creativity as Darwin's theory of evolution, Edison's lightbulb, van Gogh's paintings of sunflowers, and Bismarck's introduction of old age pensions? This chapter will develop and evaluate the *combinatorial conjecture* that all creativity, including scientific discovery and technological invention, results from combinations of mental representations.

The combinatorial conjecture has been proposed or assumed by many authors, but the evidence presented for it has been restricted to a few examples (e.g., Boden, 2004; Finke, Ward & Smith, 1992; Koestler, 1967; Mednick, 1962; Poincaré, 1913; Thagard, 1988, 1997). This chapter gives a more thorough evaluation of the conjecture by seeing whether it applies to 100 important cases of scientific discovery and to 100 important cases of technological invention. The primary result of examination of these cases is support for the combinatorial conjecture: No counterexamples were found. But the study of 200 creative episodes was interesting in other ways, and this chapter will report a collection of findings about the nature of the representations and processes used. These findings concern the role of visual and other kinds of representations and the extent to which the discoveries and inventions looked at were accidental, analogical, and observational or theoretical.

Chapter 8 used convolution in neural networks to provide a theoretical perspective on creative conceptual combination by means of a new neurocomputational mechanistic explanation of how spiking neural representations can be combined. This account accommodates visual and other nonverbal kinds of representations and therefore is capable of applying to a wide range of creative episodes. I then describe the results of study 1, which looks at 100 examples of scientific discovery, and study 2, concerning 100 examples of technological invention. These large samples confirm the combinatorial conjecture, whose claim, however, is nontrivial, as I will show by considering theoretical objections to it from the extreme embodiment perspective that thinking, and hence creativity, is not representational and computational. I will argue that these objections are unwarranted and that the combinatorial conjecture remains highly plausible for scientific discovery and technological invention.

If creativity is to be explained as combination of mental representations, we need a rigorous scientific account of the nature of representations and the processes that combine them. Chapter 8 used a neurocomputational model to show how representations construed as brain processes can be combined. Accordingly, the combinatorial conjecture can be fleshed out as follows: All creativity results from convolution-based combination of mental representations consisting of patterns of firing in neural populations. The historical studies to be described next do not serve to evaluate these neurocomputational claims, but directly address the underlying assumption that representation combination is the fundamental mechanism of creativity in various domains.

Study 1: Scientific Discovery

Case studies in the history, philosophy, and psychology of science have usefully looked in detail at select examples of advances in science and technology (e.g., Gorman et al., 2005). However, generalizations about the nature of scientific discovery need a more systematic look at a large number of episodes. Accordingly, I conducted an analysis of the cases described in a book called *100 Greatest Science Discoveries of All Time* (Haven, 2007). The author, Kendall Haven, is a reputable science journalist with a background as a research scientist with many publications. For my purposes, there is no need to defend the claim that these are exactly the "100 greatest," only that they are undeniably a large collection of very important discoveries, from the law of the lever to the human genome. Most crucial for a serious test of the combinatorial conjecture, the examples were not chosen by me and so were not biased by motivation to confirm rather than refute it. For each of the 100 discoveries, it was possible to identify concepts whose combination contributed to the discovery. The first example discussed by Haven is Archimedes's discovery of the principle of the lever, that the weights pushing down on each side of a lever are proportional to lengths of the board on each side of the balance point. This principle is a newly created proposition, which I take to be a mental representation analogous to a sentence. Philosophers sometimes talk of propositions as abstract meanings of sentences, but there is no reason to believe in the existence of such abstract entities, so I will employ the cognitive-science idea of a proposition as a mental representation carried out by neural processes. The new proposition about levers is clearly the result of combining other mental representations in the form of concepts such as *weight*, *push*, *side*, and *proportional*. Hence Archimedes's discovery of the principle of the lever confirms the conbinatorial conjecture, as do Haven's 99 other. The spreadsheet containing my analysis of the 100 is available on request.

The example that presented the biggest potential challenge to the combinatorial conjecture was the 1938 discovery in South Africa of a coelacanth, from a species that was thought to have been extinct for over 80 million years. The curator of a local museum came across a novel fish and sent it to a biologist who recognized it from fossil records. My first impression was that this discovery was a simple perceptual recognition that did not amount to the generation of any new representations. On reflection, however, it became clear that what made this discovery creative was recognition of the existence of coelacanths that are currently alive. The criteria for creativity, as suggested by Boden (2004), are that a development be novel, surprising, and important. Merely finding a coelacanth fossil would not satisfy any of these criteria, but a live specimen was indeed surprising and important. Thus the creative discovery in this case is not just the recognition of a coelacanth, but the proposition that living coelacanths currently exist. This proposition requires the combination of concepts such as *living* and *coelacanth*, and hence serves to confirm rather than refute the combinatorial conjecture. Similarly, the serendipitous discovery of penicillin might be erroneously construed as simply a matter of perception, but what made Fleming's discovery novel, surprising, and important was his more complex recognition that mold was killing bacteria, producing the key conceptual combination *bacteria-killing mold*. Serendipity and conceptual combination go together.

Note, however, that the coelacanth discovery did not require the generation of any new concepts, as the concept *coelacanth* was already familiar to evolutionary biologists. I was surprised to find that I could identify original new concepts (indicated by newly coined words) in only 60 of the 100 cases. This is probably an undercount that could be increased by more detailed study of the cases, but there still seem to be many cases where a major scientific discovery was made without introducing novel, permanent concepts. Combination of concepts into new propositions, but not permanent new concepts, include Copernicus on the Earth rotating around the Sun, Galileo on falling objects, and Boyle's law of gases. It is important to recognize that the combinatorial conjecture is true concerning mental representations in the form of propositions, but would be false if it were interpreted as a claim about creativity requiring the generation of new, permanent concepts.

In looking at Haven's sample of scientific discoveries, I was interested in what kinds of mental representations were used in discovery generation. Obviously, all 100 involved verbal representations that have been used to communicate them to others, but I conjectured that visual and other kinds of mental representations that encode information in nonverbal formats might also be relevant. (For a review of different kinds of mental representation, see Thagard, 2005a.) Mathematical representations are a subset of verbal representations that use numbers and/or equations. There are many scientific discoveries in which mathematics was important, ranging from Archimedes's law of the lever to Einstein's generation of $E = mc^2$. I identified 46 of the 100 discoveries as involving mathematical representations, although some digging could probably increase this number. Mathematical representations were much more common in physics than in biology and medicine, where many discoveries such as the existence of cells were qualitative. For tables summarizing the results described in this section and comparing them to those of technological invention, see the next section.

What role do nonverbal representations play in scientific creativity? Visual representations akin to pictures seemed to me important in at least 41 of the discoveries, ranging from Copernicus picturing the Earth going around the Sun to Bakker's imagining the activities of warm-blooded dinosaurs. Some of these were more obviously visual than others as seen from their pictorial presentations, for example in Vesalius's revisionary drawings of human anatomy and Hooke's drawings of cells viewed through a microscope. I found only five examples, however, where nonverbal, nonvisual representations seemed to play an important role in creative thinking, although more detailed historical analysis may well turn up more. I speculate that kinesthetic representations contributed to Archimedes's discoveries about levers and buoyancy and possibly to Galileo's discoveries about falling objects. Touch seems to have been relevant to Franklin's discoveries about lightning and electricity because he felt sparks, and to Rumford's ideas about heat from friction. Sound definitely contributed to Doppler's thinking about shifting frequencies. Otherwise, scientific thinking seems to have operated well just with visual and verbal (including mathematical) representations. The next section reports that nonverbal representations are much more important in technological invention.

Analysis of this relatively large sample of scientific discoveries provided an opportunity to examine questions independent of the combinatorial conjecture. I was curious how many of the discoveries were based in large part on accidents, that is, events that were not the result of an intentional plan of investigation. Based on Haven's brief accounts and my own knowledge of historical events, I estimated that around a quarter of the discoveries had a substantial accidental component. For example, Galileo was not looking for moons of Jupiter with his telescope, van Leeuwenhoek was not seeking microbes with his microscope, and Roentgen was very surprised to encounter X-rays. Hence serendipity is an important part of scientific discovery, but the majority of cases seem to result from intentional problem solving. For a discussion of many cases of serendipity in scientific discovery, see Roberts (1989) and Meyers (2007).

Most unintended discoveries are observational, but serendipity can also be a feature of theoretical research. Heisenberg's uncertainty principle was an unanticipated consequence of his mathematical explorations, and Lorenz was surprised to find large outcomes from tiny changes in the starting conditions of his computational model of atmospheric storms.

Analogy has often been recognized as an important creative cognitive process, and many important examples have been identified (Holyoak & Thagard, 1995, ch. 8). I found a significant analogical component in 14

Table 9.1
Scientific discoveries based on analogy. See text for explanation of the asterisk in "long*."

Discovery	Target	Source	Modality	Distance
living cells	living cells	monk cells	visual	long
gravity	planetary motion	projectile motion	visual, mathematical, kinesthetic	long*
fossils	sharks teeth	stone teeth	visual	long
life	hierarchy	tree	visual	long
lightning	lightning	spark	visual, heat	long*
vaccination	smallpox	cowpox	visual	local
ultraviolet light	ultraviolet	infrared	mathematical	local
electromagnetism	magnetism	electricity	visual, mathematical	long*
evolution	natural selection	Malthusian competition	mathematical	long
periodic table	elements	piano scale	visual	long
relativity	gravity	elevator	visual, mathematical	long
fault lines	rock layers	rubber bands	visual	long
Earth mantle	Earth	egg	visual	long
quantum theory	electrons	crystals	mathematical	long

cases of Haven's sample, with many cases that have not been analyzed in the philosophical or psychological literatures on analogy. It is quite possible that more examples of analogy could be found through more detailed historical analysis of the cases, but I doubt that would change the conclusion that analogy is an important but by no means exclusive mechanism for scientific creativity. Table 9.1 summarizes 14 cases of analogical discovery, noting the source analog that generated ideas about the target domain leading to a discovery. Table 9.1 also indicates the representational modalities used in addition to the ubiquitous verbal one. It is interesting that visual representations seem to be important in a greater proportion of analogical discoveries (11/14) than in discoveries in general (41/100).

Dunbar (1995) distinguished between *local* analogies that operate within a single domain and *long-distance* analogies that cross domains. All but two of the important analogical discoveries are long-distance, requiring a major mental leap across domains. Particularly interesting are three cases, indicated with an asterisk in "long*," where the analogy served to unite previously disparate domains. Before Newton, projectile and planetary motion were distinct domains, but Newtonian mechanics unified them in a common physical framework. Similarly, before Franklin, sparks and lightning were different kinds of things, and before Faraday electricity and magnetism were unconnected. In these three cases, analogical thinking brought about an important kind of conceptual change in the nature of domains that were changed through new unifying theories, so that a long-distance analogy turned into a more local one.

Philosophers often debate about the relative importance of theory and observation in science, so I coded the 100 examples for whether the discoveries were primarily:

observational, based on perception using human senses;

instrumental, based on observations using instruments; or

theoretical, requiring hypotheses that go beyond the results of sensory and instrumental observation.

The results are interesting, with 70 of the discoveries theoretical, 18 observational, and 12 instrumental. Examples of discoveries made with unaided observations include Davy's discovery of anesthesia, Mendel's findings about heredity, and Fleming's discovery of penicillin. Instruments important for making observations not possible with ordinary human perception include the telescope (Galileo), microscope (Pasteur), and spectograph (Hubble).

Finally, I was interested in the extent to which scientific discoveries depended on previous technological advances and the extent to which science led to subsequent advances in technology. My counts are preliminary and should be viewed only as approximate minimums to be made more precise by more thorough historical analysis, but they are nevertheless interesting. I found 37 of the discoveries to depend in important ways on technological advances. This was obviously much more than the 12 that were directly based on instrumental observations: many of the theoretical discoveries arose from observations that required new technologies, for example, Davy's electrochemical ideas resulting from the availability of batteries. I identified 21 scientific discoveries that in turn led to technological advances, such as Franklin's discoveries about electricity enabling him to invent the lightning rod. Thorough historical research would undoubtedly generate more examples.

Thus the study of 100 examples of scientific discovery was useful for much more than just testing the combinatorial conjecture. It served to clarify the differences between generating new propositions and generating new concepts, with only the former occurring in all cases of scientific discovery. Verbal representations seem to be universal in discoveries, but are complemented in many cases by mathematical and visual ones. Other sensory representations did not seem to be very important for scientific discovery. For lack of data, I have not addressed the role of another kind of nonverbal representation that is important for human thinking— emotion (see Thagard, 2006a, 2010a). I would conjecture that every one of the 100 discoveries generated a strong, positive emotional response in the discoverer, just like Archimedes's famous *Eureka!* moment when he discovered the principle of buoyancy. Chapter 8 gave an account of the neural processes that might generate such treasured moments.

This study has also generated interesting data about the extent to which discovery is accidental, analogical, theoretical, observational, and dependent on instruments. These data must be viewed as highly tentative, since they are based on brief accounts of one author and background knowledge of one interpreter. I hope they will serve to stimulate further systematic research on larger samples of scientific discoveries that are examined in much more depth. Let us now compare scientific discovery with a similar survey of cases of scientific invention.

Study 2: Technological Invention

Previous investigations based on a few examples have suggested that invention of new technologies involves the same basic set of cognitive processes as scientific discovery (Thagard & Croft, 1999). Both require basic cognitive processes such as problem solving, analogical inference, and concept generation. But a study of a large number of inventions also turned up some interesting cognitive differences.

For my sample of inventions, I used *100 Greatest Inventions of All Time* by an experienced nonfiction writer, Tom Philbin (2003). As with my discovery sample, the "greatest" assertion should not be taken too seriously, but there is no question that Philbin identified many very important inventions, ranging from the wheel (no. 1) to the video recorder (no. 100). Unlike Haven, Philbin ranks the creations 1 through 100 in order of

Table 9.2
Kinds of representation used in scientific discovery and technological invention.

	Verbal representation	Visual representation	Mathematical representation	Other kind of representation
Scientific discovery	100	41	46	5
Technological invention	100	87	26	48

importance, but it would be hard to defend his entire ordering. A few examples appear on both lists: anesthesia, X-rays, and the transistor. As with the 100 discoveries, the analysis of 100 inventions should be viewed as highly provisional, since it may depend on idiosyncrasies of both Philbin and myself. Still, preliminary data may help to point the way to future studies that are broader and deeper.

Tables 9.2–9.5 summarize the similarities and differences found between discovery and invention. As table 9.2 shows, all inventions, like all discoveries, involved verbal representations. This unanimity may be an artifact of the need for people (including the creators as well as commentators such as Philbin) to use language to report their discoveries to others, but inspection suggests language may well have played a role in all cases. For example, the invention of the wheel plausibly had a substantial nonverbal component owing to visual and kinesthetic representations of crucial ingredient concepts such as *log* and *rolling*, but these concepts also have verbal representations as well.

It is striking in table 9.2 that visual representations seem to be much more common in the sample of inventions than in the sample of discoveries. The reason for this difference is probably that most inventions are *things* that people can see: the wheel, lightbulb, computer, and so on. Hence people naturally have visual representations of them. For the same reason, inventions are more susceptible to other nonverbal representations such as touch, heat, kinesthesia, and sound. Mathematical representations were much less common for discoveries than for inventions, 26 rather than 46. This discrepancy may reflect the fact that Philbin had many more early examples than Haven, including ten inventions before the Christian era in contrast to only one discovery that early. Cases such as the plow, sail, and bow and arrow are clearly important inventions, but the frequency of

Table 9.3
Kinds of novelty in scientific discovery and technological invention.

	New propositions	New concepts	Accidental	Analogical
Scientific discovery	100	60	26	14
Technological invention	100	100	5	12

mathematical representations would have increased if Philbin had included many more recent science-dependent examples.

Table 9.3 summarizes aspects of novelty in discovery versus invention. All invention, like all discovery, generates new propositions, minimally of the form: This device serves to perform that function. Unlike discoveries, many of which do not introduce new concepts, all the inventions involved the introduction of new concepts. The difference again arises from the fact that all inventions are things, and the selected kinds were clearly important enough to warrant naming. Inventions seem to have occurred much less accidentally than discoveries, 5 versus 26, since inventions are usually the result of an intentional effort to solve some identified problem. Nevertheless, a few interesting cases of inventions, such as anesthesia and X-ray machines, arose accidentally: Davy was not looking for a way to kill pain, and Roentgen was not looking for a way to examine bones. I was struck, however, by the highly incremental nature of invention, with many new technologies being part of a whole series of improvements in attempts to accomplish some task such as building a better lightbulb. According to Philbin's descriptions, 71 of the inventions were incremental in this way, whereas the scientific discoveries seemed to involve more dramatic leaps.

It would be interesting to determine by more detailed historical examination whether these incremental examinations can be viewed as cases of analogical inference, using past inadequate inventions as source analogs to develop new, improved targets. That finding would increase dramatically the occurrence of analogies in invention, which at 12 percent is a bit less than the 14 percent for discovery. Table 9.4 displays the analogies identified for technological invention. As with discovery, analogy seems to have been an important cognitive process in invention, but it is far from

Table 9.4
Analogies in technological inventions.

Invention	Target	Source	Modality	Distance
printing press	printing	olive press	visual, kinesthetic	long
telephone	telephone	ear	visual, sound	long
paper	bark paper	hemp paper	visual, touch	local
airplane	airplane	bird	visual	long*
stethoscope	stethoscope	wood	sound	long
microscope	microscope	eyeglasses	visual	local
Braille	Braille	previous dots	touch	local
incubator	baby incubator	chick hatchery	visual, heat	long*
cotton gin	cotton machine	hand movements	visual, kinesthetic	local
windmill	windmill	sail	visual	long
washing machine	machine	hand movements	visual, kinesthetic	local
oil derrick	oil	gallows	visual	long

Table 9.5
Theory and observation in discovery and invention.

	Theoretical	Observational	Instrumental
Scientific discovery	70	18	12
Technological invention	24	76	0

being universal. All the inventive analogies plausibly have a nonverbal component. There are proportionately more local analogies than I found in discovery (5/12 vs. 2/14). Two of the long-distance analogies indicated by an asterisk were originally cross-domain but redefined the nature of domains so that we can now view them as the same domain. For example, before the Wright brothers used what they knew about birds to inform airplane construction, birds and flying machines were different kinds of objects, but now are unified under the general theory and practice of aerodynamics.

Finally, table 9.5 shows an interesting difference between the theoretical and observational status of discoveries and inventions. I counted 76 of the inventions as observational in that they were made using ordinary human senses, with only 24 requiring theoretical leaps beyond observation. Many of these theory-based inventions were electrical devices such as the telephone. Instruments for measuring the effects of theoretical entities were undoubtedly important in many of these inventions, but none seemed to be based just on instrumental observations without a large theoretical contribution.

Comparison of discovery and invention raises interesting questions about the relation of science to technology. I found 37 cases of scientific discoveries that depended on preceding technological developments such as the telescope and spectograph. Moreover, at least 21 of the discoveries led to new technologies such as radio and the atomic bomb. Surprisingly, looking at inventions I only identified nine of Philbin's cases as depending on prior scientific discoveries, and only three as generating new scientific discoveries. Perhaps these low numbers result from Philbin's assessment of "greatest" in terms of everyday usage rather than scientific importance. A sample of twentieth-century inventions would probably display a much stronger interconnection of technology and science more in accord with the findings from Haven's 100 discoveries.

Objections to Combination

My survey of 200 examples of scientific discovery and technological invention did not turn up any counterexamples to the combinatorial conjecture, but there are other ways in which that conjecture might turn out to be false. This section considers three: abstraction, mutation from single representations, and (most radically) creativity that does not at all rely on mental representations. I have already argued that serendipitous perception in cases such as the coelacanth and penicillin are actually cases that confirm rather than refute the combinatorial conjecture.

Welling (2007) discusses four creative processes: application of existing knowledge, analogy, combination of concepts, and abstraction. The first three of these clearly involve combinations of representations, but what about the fourth, abstraction? According to Welling (2007, p. 170):

The mental *process* of abstraction may be defined as: the discovery of any structure, regularity, pattern or organization that is present in a number of different perceptions that can be either physical or mental in nature. From this detection results the *product* abstraction: a conceptual entity, which defines the relationship between the elements it refers to on a lower, more concrete, level of abstraction.

To illustrate abstraction, Welling uses Piaget's example of children learning the concept of *weight* by abstraction of experiences of objects that are heavy and light. He speculates that children and other learners use Gestalt principles of perceptual organization such as grouping and closure.

Without a more detailed model of how abstraction works, it is difficult to assess whether it constitutes a challenge to the combinatorial conjecture. It does seem, however, that abstraction requires the combination of perceptual representations, for example, the physical, kinesthetic sensations involved in assessing an object as heavy or light. I conjecture that when children learn the abstraction *weight* they put together a combination of verbal and nonverbal representations of experiences of heavy objects and light objects. If there are any cases of abstraction that are not combinatorial in this fashion, I expect that they are not particularly creative according to Boden's criteria of being new, surprising, and valuable.

Another kind of possible counterexample to the combinatorial conjecture would be if creativity arose from a kind of mutation in a single concept, analogous to the way that mutations occur in genes. Although various authors have attempted to exploit an analogy between genetic mutation and concept generation (e.g., Dawkins, 1976), I think this analogy is feeble from a cognitive perspective (Thagard, 1988, ch. 6). Thinking is much more structured and constrained than biological evolution. In particular, no one has ever identified an interesting case of creativity arising from a random alteration in a single concept analogous to mutation in a single gene. Conceptual combinations occur in a much more focused way in the context of problem solving and hence are a much more plausible mechanism of creativity than single-concept mutations. See the appendix to this chapter for further analysis of why discovery and invention do not result from blind variations.

My analyses and arguments in defense of the combinatorial conjecture may suggest to the reader that the claim is true but trivial, making no substantive assertion. Response to the charge of triviality can be made first by pointing to the neurocomputational theory of combination that proposes a detailed neural mechanism for combining representations using convolution (chapter 8). This theory shows that the conjecture can be fleshed out into a specific claim about the cognitive and emotional processes that underlie human creativity. Second, the nontriviality of the combinatorial conjecture is evident from serious theories that deny a theoretical role for representation altogether. If there are no mental representations, then creativity is obviously not the result of combining them. Denial of mental representations was a hallmark of behaviorism, which dominated American psychology until the cognitive revolution of the 1950s. It has been revived in a movement espousing "radical embodied cognitive science" that draws on a combination of Heideggerian philosophy, Gibsonian psychology, and dynamic systems theory to propose an alternative to the dominant computational-representational view of thinking (see, e.g., Chemero, 2009; Clark, 1997; Dreyfus, 2007; Thompson, 2007; Warren, 2006). If radical embodiment is true, then creativity does not require combining representations at all: it can be "action-first" rather than "thought-first" (Carruthers, 2011). Then the combinatorial conjecture would be false.

These extreme embodiment claims need to be distinguished from more moderate ones made by researchers such as Gibbs (2006) and Barsalou et al. (2003), who maintain that the kinds of representations and computations performed by the brain are closely tied to bodily processes such as perception and emotion. My account of neural representation is sufficiently broad to encompass a wide range of perceptual modalities, all of which can be understood as patterns of activation in populations of neurons. The body also plays a large role in my account of emotion (Thagard, 2010a; Thagard & Aubie, 2008). Hence I am happy to endorse a *moderate* embodiment thesis that acknowledges the importance of perceptual and other physiological processes (chapter 4). This moderate thesis is fully compatible with the combinatorial conjecture as long as cognition is not mistakenly restricted to only languagelike representations.

So why does cognitive science, and particularly the theory of creativity, need representations? The answer is that postulation of various kinds of representations currently provides the best available explanation of many kinds of human thinking, including perception, inference, learning, problem solving, and language use (see, e.g., Anderson, 2010; Smith & Kosslyn, 2007; Thagard, 2005a). Proponents of the extreme embodiment thesis have barely scratched the surface in matching the explanatory successes of the computational-representational approach. Indeed, chapter 4 argued that even the basic problem of motor control is too complex to understand without postulating representations and computations

(Todorov & Jordan, 2002; Wolpert & Ghahramani, 2000). Abilities such as grasping objects require building complex mental models to predict the effects of various kinds of muscular operations.

More specifically, no one has a clue how to use pure embodiment to explain creative developments in science and technology. Discoveries such as relativity theory and inventions such as the telephone require the full range of human representational capacities, from verbal and mathematical to more obviously embodied representations such as vision and sound. Humans are indeed embodied dynamic systems embedded in their environments, but our success in those environments depends heavily on our ability to represent them mentally and to perform computations on those representations. Hence the embodied aspect of much of mental cognition does not refute the combinatorial conjecture, although the claims of radical embodiment do serve to show that the conjecture is a substantive one about human creativity.

According to Arthur (2009), new technologies arise as combinations of other technologies. Obviously, past technologies have not had the capability to actually combine or modify themselves, although future, more intelligent machines may do so (Lipson & Pollock, 2000). Rather, new technologies from wheels to iPads have resulted from the combination of human mental *representations* of previous technologies. Technological creativity is both a physical process of interaction with the world and a social process of interaction with other people. But it is also a psychological process carried out by brains that are capable of computationally modifying representations through such mechanisms as visualization, conceptual combination, analogy, and inference in general.

Conclusion

The two studies in this chapter have found support for the combinatorial conjecture in 200 examples of discovery and invention, but do not address whether it holds in other domains of human creativity. It should not be too hard to apply the conjecture to social innovation, which concerns the creation of new organizations, institutions, and practices that benefit human society. Innovations such as democratic government, public education, pension plans, universal health care, and international governance have contributed greatly to the quality of human lives, and I expect to show that all of these resulted in part from the combination of representations. My guess is that social innovation will turn out to be more like technological invention than like scientific discovery, except for a reduced contribution of representations that are visual.

More difficult will be the assessment of the applicability of the combinatorial conjecture to the great many examples of creativity resulting from artistic imagination, including poetry, plays, novels, films, music, dance, and architecture. Examination of even a few examples from these categories will require attention to many kinds of representation beyond the verbal, such as emotion in poetry, vision in film, sound in music, and kinesthesia in dance and sculpture. I expect that scrutiny of such examples will serve not just to confirm the combinatorial conjecture but also to flesh it out with greater understanding of the kinds of representations and processes that contribute to human creativity. Also needed is a general theory of how newly generated representations are evaluated for their coherence with other representations and overall value.

Although my analysis of 200 examples has been highly provisional and needs to be supplemented by a deeper and broader study, it has helped to characterize representational aspects of creativity along such dimensions as mode of representation, role of accident and analogy, and relative contribution of theory and observation. I would like to see the development of an *Atlas of Creative Science and Technology* that would contain not only historical descriptions of great discoveries and inventions but also their assessment with respect to the kinds of cognitive factors identified in this chapter. Creativity is combination of representations, but there is much more to be learned about the nature of these representations and the cognitive processes that produce them.

Appendix: Blind Variation

Simonton (2010) has attempted to revive the idea of blind variation in creativity, but I think his mathematical analysis does not go to the heart of the matter. Here is an alternative.

Let V be the set of variants that can arise in a set of structures such as genes, mental representations, machines, etc. V will be very large, but not infinite, as biological and physical entities are finite. The variants can be numbered $V_1 \ldots V_k$, with V_i indicating some specific variant.

Then we can define G_i as the generation of variant V_i, and U_i as the utility of V_i. I propose that a variant is *blind* if its generation is independent of its utility, i.e., the probability of generation given nonzero utility is the same as the probability of its generation if it were useless:

V_i is blind iff $P(G_i / U_i > 0) = P(G_i / U_i = 0)$.

Then we can say that a process of variation is overall-blind if every variant in it is blind. Genetic mutation is overall-blind, but scientific discovery, technological invention, and other forms of human creativity are overwhelmingly not, because psychological processes of problem solving and representation generation focus thinking toward variants that are more useful than random ones.

A quantitative approach has the advantage that we can talk about degree of blindedness, which is the extent to which variants in a process are blind:

Blindedness = no. of blind actual variants / no. of actual variants.

My survey of 100 scientific discoveries and 100 inventions suggests that the blindedness of these processes is near zero, although the difficulty of assessing the relevant probabilities and utilities makes it hard to say. Many discoveries (but hardly any inventions) have an unintentional component, but even in these cases it seems that more useful variants are more likely to be generated than useless ones. For example, Galileo never intended to find the moons of Jupiter when he turned his new telescope on the heavens, but his interests, background knowledge, and cognitive processes made it more probable that he would generate the representation "Jupiter has moons" than some utterly useless representation such as "Rome has toes." Hence, discovery is not blind, and biological evolution is a poor model for scientific discovery and other kinds of creativity.

10 Creativity in Computer Science

Daniel Saunders and Paul Thagard

Introduction

Computer science only became established as a field in the 1950s, growing out of theoretical and practical research begun in the previous two decades. The field has exhibited immense creativity, ranging from innovative hardware such as the early mainframes to software breakthroughs such as programming languages and the Internet. Martin Gardner worried that "it would be a sad day if human beings, adjusting to the Computer Revolution, became so intellectually lazy that they lost their power of creative thinking" (Gardner, 1978, pp. vi–viii). On the contrary, computers and the theory of computation have provided great opportunities for creative work.

This chapter examines several key aspects of creativity in computer science, beginning with the question of how problems arise in computer science. We then discuss the use of analogies in solving key problems in the history of computer science. Our discussion in these sections is based on historical examples, but the following sections discuss the nature of creativity using information from a contemporary source, a set of interviews with practicing computer scientists collected by the Association of Computing Machinery's online student magazine, *Crossroads*. We then provide a general comparison of creativity in computer science and in the natural sciences.

Nature and Origins of Problems in Computer Science

Computer science is closely related to both mathematics and engineering. It resembles engineering in that it is often concerned with building machines and making design decisions about complex interactive systems. Brian K. Reid wrote: "Computer science is the first engineering discipline ever in which the complexity of the objects created is limited by the skill of the creator and not limited by the strength of the raw materials" (Frenkel, 1987, p. 823). Like engineers, computer scientists draw on a collection of techniques to construct a solution to a particular problem, with the creativity consisting in the development of new techniques. For example, during the creation of the first large-scale electronic computer, Eckert and Mauchly solved numerous engineering problems, resulting in solutions that became important contributions to computer science (Goldstine, 1972).

Computer science also has a strong mathematical component. An early example is Alan Turing's invention of an abstract, theoretical computing machine in 1935 to solve David Hilbert's decidability problem in the foundations of mathematics (Hodges, 1983). The Turing machine has proven to be a very powerful tool for studying the theoretical limitations of computers. The theory of the class of NP-complete problems, which appear to be computationally intractable, is another result in both mathematics and computer science (Cook, 1971; Garey & Johnson, 1979).

Computer science is like engineering in that it is often concerned with questions of how to accomplish some technological task rather than with scientific questions of why some natural phenomenon occurs. Like mathematics, computer science is largely concerned with manipulating abstract symbols. In contrast to natural science, stories of serendipitous discovery are uncommon: computers often do unexpected things, but rarely in a way that leads to new discoveries. In a company, problems can come from commercial motivations such as the desire to enter into a new market, to improve on an existing product, or to overcome some difficulties that are preventing a project from being delivered. In the academic world, questions often arise from reading already published papers that mention extant problems.

There is in addition a source of problems that is especially important in computer science. Computers are unusual machines in being not only tools but tools for making tools. What computer scientists study is also a machine that can help with its own investigation, as if microbiologists studied their electron microscopes rather than bacteria. In a typical scenario, a computer scientist becomes frustrated with a repetitive, boring, and difficult task that might be relieved by new technology. When someone with the right background and ideas experiences such frustration, creative contributions to computer science can result.

A major example occurred during the final months of work on the ENIAC, the first general purpose electronic computer, started at the Moore School of Electrical Engineering in 1943 (Goldstine, 1972). ENIAC had been built with the assumption that only one problem would be run on it for a long period of time, so it was difficult and time-consuming to reprogram; in essence it had to be rewired by physically plugging in wires to make a new circuit. This led to great frustration, especially when the team's machine was pitted against the much slower and more primitive Harvard-IBM Mark I machine, which did not use vacuum tubes: "To evaluate seven terms of a power series took 15 minutes on the Harvard device of which 3 minutes was set-up time, whereas it will take at least 15 minutes to set up ENIAC and about 1 second to do the computing" (Goldstine, 1972, pp. 198–199).

The solution to this problem, primarily due to John von Neumann, was the concept of a stored program, which enabled the instructions for the computer to perform in the same memory as the data on which the instructions were to operate. With stored programs, a computer could be set to work on a new problem simply by feeding it a new set of instructions. The key insight was that instructions could be treated just the same as data in an early memory device in which numbers were stored as sound waves echoing back and forth in tubes of mercury.

Another example of frustration-based creativity is the invention of the first high-level language, FORTRAN, by John Backus in 1953, which allowed computer programs to be written in comprehensible, algebra-like commands instead of assembly code (Shasha, 1995, pp. 5–20). Assembly code consists of lists of thousands of inscrutable three-letter commands, designed to communicate with the machine at its lowest level. Backus was a programmer at IBM, who said later: "Assembly language was time-consuming; it was an arduous task to write commands that were specific to a particular machine and then to have to debug the multitude of errors that resulted" (ibid., p. 10). The layer of abstraction that FORTRAN placed between the underlying electronics and the human was very important to the subsequent development of computer science. Programs could be written faster and with fewer errors, making possible larger and more complex uses of the computer, including building yet more powerful tools and programming languages.

There are many other examples of creativity inspired by frustration in computer science, such as the Unix operating system, Larry Wall's programming language Perl, and Donald Knuth's (1974) typesetting system T_EX. Larry Wall declared that the three virtues of the computer programmer are "laziness, impatience, and hubris" (Wall, 1996, p. xiii). It is thus clear that frustration with the limitations of a machine or the repetitive nature of a task is an important motivating force in producing creative answers in computer science. For a survey of important innovations in computer software, see Wheeler (2010).

Although technological frustration may be the origin of many problems in computer science, we should not neglect the intrinsic pleasure for many people of building computers and writing computer programs. According to Brooks (1982, p. 7): "The programmer, like the poet, works only slightly removed from pure thought-stuff. He builds castles in the air, from air, creating by exertion of the imagination. Few media of creation are so flexible, so easy to polish and rework, so readily capable of realizing grand conceptual structures." The term "hacker" referred originally not to people who use programs for destructive purposes such as breaking into the computers of others, but to creative young programmers who reveled in producing clever and useful programs (Levy, 1984; Raymond, 1991). Presumably, there are also computer scientists whose creativity is fueled by financial interests: excitement can derive from the prospect that novel software and hardware can make the inventor rich.

Not all computer scientists are concerned with practical problems. The theory of NP-completeness originated because of Stephen Cook's interest in mathematical logic, although he was also interested in computational implications (Cook, personal communication, July 4, 2002): "My advisor at Harvard, Hao Wang, was doing research on the decision problem for predicate calculus. Validity in the predicate calculus is r.e. complete, and this was what gave me the idea that satisfiability in the propositional calculus might be complete in some sense." Thus Cook formulated a new problem in theoretical computer science by analogy to a familiar problem in mathematical logic. Analogies can also be a fertile source of solutions to problems.

Creative Analogies in Computer Science

Once computer scientists have posed a problem, how do they solve it? Creative problem solving about computers requires all the cognitive processes that go into scientific research, including means-ends reasoning with rules, hypothesis formation, and generation of new concepts. This section will focus on one important source of creative problem solving, the use of analogies. Analogy is far from being the only source of creative solutions in computer science, but its importance is illustrated by many historical examples.

Following Dunbar (2001) and Weisberg (1993), we distinguish between *local* analogies, which relate problems from the same domain or from very similar domains, and *distant* analogies, which relate problems from different domains. Distant leaps from one domain to another have contributed to important scientific discoveries such as Benjamin Franklin's realization that lightning is a form of electricity (Holyoak & Thagard, 1995, ch. 8). Distant analogies have also played an important role in the history of computer science, but local analogies have also been important for creativity in computer science.

An early distant analogy in computer science was made by Charles Babbage when he realized that the punch cards that controlled the intricate weaving on the automatic Jacquard looms of the day might be used with his hypothetical computing machine, the Analytical Engine (Goldstine, 1972, ch. 2). Lady Ada Lovelace, his friend and collaborator, wrote, "We may say most aptly that the Analytical Engine weaves algebraic patterns just as the Jacquard-loom weaves flowers and leaves" (Goldstine, 1972, p. 22).

Biological analogies have been particularly important in computer science: von Neumann in 1945 was influenced by McCulloch and Pitts's theories of how the human nervous system could be described mathematically, and he used their notation in his "First Draft of a Report on the EDVAC" (Goldstine, 1972, pp. 196–197). Another example is the invention by Bob Baran of the packet-switching network, the technology underlying the Internet (Hafner & Lyon, 1996). In the context of the nuclear fears of the time, he wanted to build a network that could survive with numerous nodes knocked out. He took for his inspiration the structure of the human brain with its redundancy and decentralization. The massively parallel

Connection Machine of Danny Hillis was built to emulate the parallelism in the brain (Shasha, 1995, pp. 188–206). More recently, genetic algorithms and neural networks have proved to be powerful biologically based ideas that are gradually being incorporated into the mainstream of computer science.

The Wright brothers, in building the first manned, powered flying machine, also used a biological analogy, modeling the turning system of their aircraft after that used by birds (Weisberg, 1993, pp. 132–148). However, the Wright brothers also drew on decades of work on flight, carefully studying analogues close to their goal such as sophisticated gliders and attempts at powered flights that had preceded their attempt. The same phenomenon is also found in computer science, where original ideas are often variations and improvements on existing technologies that furnish local analogies.

Eckert and Mauchly had a local inspiration, in the form of John Atanasoff's ABC computer (Slater, 1987, pp. 59–63). The ABC was a special-purpose machine, designed only to solve differential equations, and there is some argument about whether it was fully operational, but it was the first of its kind, using vacuum tubes to perform its operations. Mauchly was keenly interested in the machine and came to visit Atanasoff for four days, talking computers with him the whole time. Only after the mountain of publicity on the ENIAC's unveiling in 1946 did Atanasoff realize how many ideas they had borrowed from his machine. In 1975, at the end of a drawn-out lawsuit sparked by the pair's attempts to defend patents on their invention, Atanasoff was declared the legal inventor of the automatic electronic digital computer.

Whether or not this was a case of intellectual plagiarism, it is typical of how inventors absorb technological ideas and build on them. FORTRAN was preceded by Speedcoding and A-0, and Babbage was familiar with the machines of Leibniz and Pascal. Creative inventors do not work in a vacuum, but are usually very aware of other creative work going on in their field.

Similarly, in 1979 Steven Jobs and engineers from Apple Computers took a tour of the Xerox Palo Alto Research Center, where they saw menus, windows, and user-friendly word processors years before they would be available to the public. At least one member of the team has regretted the effect of this event on the perception of the creativity of the Macintosh group, saying, "Those one and a half hours tainted everything we did, and so much of what we did was original research" (Hiltzik, 1999, p. 343n). Later, Microsoft borrowed many of the same ideas for its Windows operating system, another case of problem solving with local analogies.

What then is the role of distant analogies if most important inventions have local antecedents? Consider the case of the invention of object-oriented programming by Alan Kay in the early 1970s. As a graduate student, he was exposed to a pioneering computer graphics program called Sketchpad, which used the concept of "master" drawings and "instance" drawings, and the computer language Simula. He was struck with the ideas in Simula, and it stimulated his thinking along analogical lines: "When I saw Simula, it immediately made me think of tissues . . . [it] transformed me from thinking of computers as mechanisms to thinking of them as things that you could make biological-like organisms in" (Kay, 1996b, p. 575). Kay gave this description of how the cumulative effect of the various examples he was exposed to led to the ideas that culminated in Smalltalk: "The sequence moves from 'barely seeing' a pattern several times, then noting it but not perceiving the 'cosmic' significance, then using it operationally in several areas; then comes a 'grand rotation' in which the pattern becomes the center of a new way of thinking" (Kay, 1996a, p. 514). Many of the parts of Smalltalk can be traced back to local analogies with Simula and Sketchpad, but the general principles of object-oriented programming that were crucial to the final design of Smalltalk did not occur to Kay until he had seen the distant biological analogy. "The big flash was to see this as biological cells. I'm not sure where that flash came from but it didn't happen when I looked at Sketchpad. Simula didn't send messages either" (Shasha, 1995, p. 43). Similarly, when James Gosling invented the programming language Java, he was inspired both by local analogies, particularly to the programming language C++, and also by a distant analogy to networks of sound and light he experienced at a rock concert (Thagard & Croft, 1999).

Distant analogies in computer science are useful not only for generating new ideas, but also in communicating ideas and suggesting deeper principles underlying the accumulation of technological ideas. Notice the reliance on metaphorical names for the new entities, such as "files" and "folders," "words" and "pages" of memory, and even stranger appropriations, such as "firewall" and "zombie." Some, such as "software engineering" and "virus," began as metaphors but later expanded and conceptually deepened the source words, so that there are now more kinds of engineering and more kinds of viruses than there used to be.

Everyday Creativity

The previous sections have described some of the most important historical cases of creativity in computer science, but have ignored the day-to-day creative processes of working computer scientists in both academia and industry. For insight into this more prosaic topic, we examined surveys compiled by *Crossroads*, the online student magazine of the ACM (2002). This magazine used to include a feature called "A Day in the Life of . . ." that interviewed practicing computer scientists. We looked at their answers to two questions: "What do you do to get yourself thinking creatively?" and "What is your problem solving strategy?" The answers were often brief, and sometimes seemed to reflect different interpretations of the questions, but displayed some interesting patterns. Out of the people profiled in the "A Day in the Life of . . ." feature, we chose fifty with either an academic position in computer science or a job title of "computer scientist" in industry. All quotations without references are taken from this source.

We have not found evidence that the creativity behind famous inventions is fundamentally different from the everyday creativity of ordinary computer scientists. Not all the respondents to the *Crossroads* questions are ordinary, for they include Herbert Simon and Internet pioneer Leonard Kleinrock. One famous name in computer science, Leslie Lamport, said: "When I look back on my work, most of it seems like dumb luck—I happened to be looking at the right problem, at the right time, having the right background" (Shasha, 1995, p. 138). This quote underemphasizes the important roles of motivation and intelligence in achievement, but plausibly suggests that creativity is not a rare or supernatural gift.

Only two of the fifty examined respondents to the creativity survey mentioned analogies, and those went no further than simply "think of analogies." The others either did not use analogies in their creative work or were not aware of using them.

In the answers to the survey's creativity questions, two modes of creative work were evident, usually one or the other being emphasized in a particular person's answer. We may call them the *intense* mode and the *casual* mode, and they are illustrated by two images of the artist or scientist at work. In the intense mode, the creative individual is hunched over a desk, scribbling madly on pads of paper. In the casual mode, the researcher is lying in the bath relaxing when a solution to a laid-aside problem suddenly hits.

The intense mode is the one that most looks like work, and it is characterized by either writing on many sheets of paper or else engaging in animated conversation. This writing is very different from the work involved in communicating one's ideas. The result is not a draft of a paper, but pages covered in scrawls and doodles, showing the evidence of prolonged attacks on an idea from many directions at once: experimentation, examples, pictures, lists, restatements of the problem—anything that could help crack it open. It is most often seen as problem solving, and work in this mode feels relatively rational and systematic. Many of the respondents used systematic language like "first lay out all the options," "ask myself questions," and "create a list of the various issues," and two even provided a numbered list of the steps they take in solving a problem, such as "2. Determine the components and parameters of the problem."

For thinkers in intense mode, the pencil and paper provide a kind of feedback loop, an extension of normal human capabilities that helps to capture thoughts and preserve them beyond the span of short-term memory. Other people can also serve to record and amplify ideas. The phrases "bouncing ideas off" and "brainstorming with" were often used in the computer scientists' thoughts about their creative method. Social mechanisms to encourage such creative communication are found in both academic and corporate workplaces.

FORTRAN inventor John Backus once said: "It's wonderful to be creative in science because you can do it without clashing with people and suffering the pain of relationships" (Shasha, 1995, p. 20). But computer scientists depend on social relationships to foster their creativity as much as other researchers. Twelve of the respondents to the survey explicitly mentioned the importance of talking to people in their creative process, and, in the five years of the *Journal of the Association for Computing Machinery* from 1997 to 2002, 132 out of 164 articles (80.5%) had more than one author.

It is surprising that, for this group of experienced computer scientists, the tool of choice for focused assault on a problem is usually ordinary writing materials such as pencil and paper. "I probably do best by writing stuff down, often free-flow." "Lots of writing on paper." "Open my note-book to a blank sheet of paper." "[Problem-solving] usually takes A LOT of paper to scribble on, some is just doodling while you stare at the pieces, others is actual notes that help with solutions." Others mentioned dry-erase whiteboards, ubiquitous in every high-tech company and research institute, as critical to their creative process, sometimes even using them as a verb, as in "Whiteboard it!" Typical is a description of Xerox PARC whiteboards covered "with boxes filled with other boxes and arrows point-ing to yet more boxes with pointers across to new boxes and so on" (Hiltzik, 1999, p. 225).

Why are computers considered inadequate for the most critical part of the development of creative ideas? Part of the reason may be the distract-ing clutter of computer displays. Another reason may be the comparable robustness and portability of a notepad and pen. But Robert Smith's answer to this question suggests there may be deeper problems: "It is very difficult to use a computer when you are trying to birth new ideas. The computer limits your ideas to the 'shape' of documents you know how to create (immediately turns everything into a text document or briefing slide). Blank paper lets you develop an idea in any form you can scribble." Perhaps someday there will be computer software that incorporates such superior characteristics of plain paper as the lack of distracting features, the ease of movement between text, formulas, and drawings, and the freedom to lay out the page flexibly.

The Casual Mode of Creativity

In contrast to the intense mode, the casual mode of creative thinking usually involves inspiration striking during a break from work. Many of the *Crossroads* respondents reported that their best insights occurred not when they were diligently working at their desks, but rather in nonwork situations. Many mentioned their time running or hiking as prime creative time, and nine believed physical exercise was important for creative think-ing. One reports this strategy for creative thinking: "Go to the beach and run and run and run! Then think and jot things down quickly, saving the refinements for later." The father of computer science, Alan Turing, was an avid long-distance runner, often running the fifty kilometers between two universities. According to his biographer, Andrew Hodges, he came to his idea of the Turing machine lying in the middle of a field after a long run

(Hodges, 1983, p. 100). Other respondents mentioned hiking, bike riding, or martial arts as important to helping them get thinking creatively.

Driving was emphasized by one respondent: "I drive an hour each way to work which gives me a lot of time to contemplate and just let my mind range from topic to topic. Most of the time I spend the drive looking at the scenery and then seeing where that topic takes me." Other researchers also report that their daily drive or subway ride often constitutes their most important creative period of the day.

The shower also seems to be an excellent place for generating creative ideas. One reported: "Once I have a problem, it becomes part of me, and ideas come up mostly in non-work situations. I came up with a major idea for my thesis while in the shower." Alan Kay, the inventor of object-oriented programming and the graphical user interface, believed he got his best ideas in the shower. He even had one to himself in the basement of the Xerox PARC research facility where he made some of his most startling inventions, using it in the morning, his most productive time of the day (Hiltzik, 1999, p. 217).

Finally, several computer scientists report getting great ideas in bed, in the middle of the night or on waking. One reported her response to getting badly stuck on a problem is to go to bed and "wake up at 2 AM with the solution suddenly obvious." The original inventor of public key cryptography, J. H. Ellis, reportedly made his discovery in the middle of the night, after spending some days mulling over the problem of how to allow secure communication by two parties who had never met (Ellis, 1987).

These occasions of casual mode creativity share the following characteristics: immersion in a problem domain, absence of immediate pressure to focus on the problem, absence of distractions, mental relaxation, unstructured time, and solitude. Physical activities are an important way of reaching a relaxed state of mind where ideas that had been studied intensely before are free to combine in various ways without immediate pressure to solve a problem. This increased freedom helps to overcome problems that have become obstacles.

An important illustration of the casual mode in computer science history is the story of John Atanasoff (Slater, 1987, pp. 53–63). Atanasoff was a professor at Iowa State College and had become very concerned with the problems his graduate students were having doing calculations with the primitive analog machines available at the time. He had thought intensely about the problems of computing, but felt stuck. He went for a very fast drive in his car until he crossed the border into Illinois, where he stopped at a roadhouse and ordered a drink. "Now, I don't know why my mind worked then when it had not worked previously, but things seemed to be good and cool and quiet. . . . I would suspect that I drank two drinks perhaps, and then I realized that thoughts were coming good and I had some positive results" (Mackintosh, 1988). In the few hours he spent at the roadhouse, he formed several important ideas for his computer, including the use of binary numbers and the use of vacuum tubes instead of relays to increase the speed of calculations. Atanasoff described the feeling of having heightened creativity, of being inspired: "I remember it perfectly, as if it were yesterday. Everything came together in that one night. All of a sudden I realized that I had a power that I hadn't had before. How I felt that so strongly in my soul, I don't know. I had a power—I mean I could do things, I could move, and move with assurance" (Slater, 1987, p. 55).

The casual mode contributes to the phenomenon of sudden insight, described by Hadamard (1945, p. 21) as "those sudden enlightenments which can be called inspirations." He quoted Helmholtz: "Creative ideas . . . come mostly of a sudden, frequently after great mental exertion, in a state of mental fatigue combined with physical relaxation." Hadamard proposed that creative insights happen in four stages, which he called preparation, incubation, illumination, and "precising." But the incubation stage, where little directed thought is given to the problem, could only lead to illumination when preceded by hard work on the problem. As Louis Pasteur said, chance favors the prepared mind. Thus insight in the casual mode requires previous work in the intense mode.

According to Jack Goldman, who was head of research at Xerox and responsible for the famously creative computer science lab at the Palo Alto Research Center: "Invention can result from a flash of genius or painstaking pursuit of a technical response to an identified or perceived need—sometimes perceived only by the inventor himself" (Alexander and Smith, 1988, p. 34). Although the intense mode feels less special and creative than when ideas come out of the blue in the casual mode, it is a critical part of creative work, and responsible for many important discoveries. Creative researchers typically construct their days to give some time to both modes.

In the *Crossroads* survey, some computer scientists report getting stimulated creatively by any kind of creative work, not just work in computer science and its related disciplines. "A good film or art exhibit or any creative work done with excellence can inspire me as much as computing." "Reading someone else's creative material starts the juices flowing again." "Reading articles in *Science* magazine about biology or something else completely outside of what I usually do is also stimulating." This seems to be more than just a case of noticing distant analogies between whatever one is reading and one's own work. Why should creative work in one domain inspire creativity in a researcher in a totally different domain? Perhaps inspiration derives from stimulation of excitement that facilitates work in general; Isen (1993) reviews research that shows that induction of positive mood enhances creative decision making.

Comparison with Natural Science

We now compare what we have learned about creativity in computer science with aspects of creativity in the natural sciences. Problems in physics, chemistry, and biology can take the form of several different kinds of questions. Some questions are empirical, inquiring about the relations between two or more variables with observable values. On the more theoretical side, there are why-questions that ask for an explanation of observed empirical relations. Sciences such as biology often also ask how-questions that can be answered by specifying a mechanism that shows how observed functions are performed. Finally, applied science such as biomedicine asks how-questions concerning how what is known about mechanisms might be used to solve some practical problem such as treating disease. Computer science is not concerned with empirical questions involving naturally observed phenomena, nor with theoretical why-questions aimed at providing explanations of such phenomena. Nor does it study naturally occurring mechanisms; rather, it usually aims at producing new mechanisms, both hardware and software, that can provide solutions to practical problems. This is computer science as engineering. The problems in theoretical computer science are more like problems in pure and applied mathematics than like problems in the natural sciences. They usually involve very abstract notions of computation and can often be only tangentially relevant to practical concerns of computing.

In the natural sciences, the origins of problems are often empirical, when surprising observations lead to experimentation and theorizing aimed at replacing surprise with understanding. Some problems arise for conceptual reasons, for example, when two plausible theories are incompatible with each other. The motivation for theoretical computer science is largely conceptual, but much work in computer science originates with practical frustrations with available technology. Thus the origins of problems in most of computer science are like the origins of problems in applied sciences such as biomedicine. Many emotions, including frustration, need, surprise, interest, and ambition can motivate novel work in both science and engineering (Thagard, 2006a, ch. 10). We said above that there seems to be less serendipity in computer science than in natural science, so perhaps frustration is a more important emotion than surprise for technological innovation.

How does the finding of solutions to problems in computer science compare with finding solutions to the empirical and theoretical problems in the natural sciences? Thagard and Croft (1999, p. 134) compared scientific discovery with technological innovation, and concluded:

Scientists and inventors ask different kinds of questions. Because scientists are largely concerned with identifying and explaining phenomena, they generate questions such as:

Why did X happen? What is Y? How could W cause Z?

In contrast, inventors have more practical goals that lead them to ask questions of the form:

How can X be accomplished? What can Y be used to do? How can W be used to do Z?

Despite the differences in the form of the questions asked by scientists from the form of the questions asked by inventors, there is no reason to believe that the cognitive processes underlying questioning in the two contexts are fundamentally different. Scientists encounter a puzzling X and try to explain it; inventors identify a desirable X and try to produce it.

In computer science the desirable X is a better way of computing. Like all problem solvers, computer scientists use means-ends reasoning and analogies to generate solutions to problems. Hence creativity in computer science seems very similar cognitively to creativity in thinking in the natural sciences.

We are not aware of any systematic studies of the roles of intense and casual modes of thinking the natural sciences. Cognitive psychologists have primarily investigated the intense mode, but there is anecdotal evidence that the casual mode contributes to creativity in the natural sciences. For example, Kekulé reported apprehending the structure of benzene during a dream. It would be interesting to conduct a survey, analogous to the *Crossroads* interviews with computer scientists, to determine whether other kinds of thinkers are also sometimes inspired in casual contexts. It would not be surprising if natural scientists also can benefit from physical exercise, showers, concerts, and lying in bed. (Thagard got one of his best ideas, the computational theory of explanatory coherence, while watching a boring movie, *Beverly Hills Cop 2*.)

In sum, the cognitive and social processes that foster creativity in computer science do not seem to be different in kind from the processes that underlie success in the natural sciences, especially applied sciences such as medicine. Computer science is certainly different in that it studies an artifact, computers and computation, rather than naturally occurring phenomena. But all creativity in science, technology, and mathematics involves asking novel questions to pose hard problems and then answering them by using analogies and other kinds of reasoning to form new hypotheses and concepts.

Conclusion

We conclude by summarizing the major findings of our investigation of creativity in computer science. Problems in computer science originate in both engineering and mathematical contexts. Many engineering problems in computer science arise from frustration with available computational tools, but creativity can also be fueled by financial interests and the pleasure of building computers and writing software. Both local and distant analogies can be useful in solving problems. Everyday creativity in computer science operates in both an intense mode of focused concentration and a casual mode in which inspiration strikes during nonwork activities. Answering computational questions seems to involve the same kinds of cognitive processes that generate answers to empirical questions in the natural sciences. Creativity requires caring about a problem sufficiently to work on it intensely, using analogies and other means of generating novel solutions.

11 Patterns of Medical Discovery

Introduction

Here are some of the most important discoveries in the history of medicine: blood circulation (1620s), vaccination, (1790s), anesthesia (1840s), germ theory (1860s), X-rays (1895), vitamins (early 1900s), antibiotics (1920s–1930s), insulin (1920s), and oncogenes (1970s). This list is highly varied, as it includes basic medical knowledge such as Harvey's account of how the heart pumps blood, hypotheses about the causes of disease such as the germ theory, ideas about the treatments of diseases such as antibiotics, and medical instruments such as X-ray machines. The philosophy and cognitive science of medicine should be able to contribute to our understanding of the nature of discoveries such as these.

The aim of this chapter is to identify patterns of discovery that illuminate some of the most important developments in the history of medicine. I have used a variety of sources to identify forty great medical discoveries (Adler, 2004; Friedman & Friedland, 1998; Science Channel, 2006; Straus & Straus, 2006). After providing a taxonomy of medical breakthroughs, I discuss whether there is a logic of discovery and argue that the patterns of medical discovery do not belong to formal logic. In contrast, it is possible to identify important psychological patterns of medical discovery by which new hypotheses and concepts originate. In accord with recent developments in cognitive science, I also investigate the possibility of identifying neural patterns of discovery. Finally, I discuss the role that computers are currently playing in medical discovery.

Medical Hypotheses

At least four different kinds of hypotheses are employed in medical discovery: hypotheses about basic biological processes relevant to health; hypotheses about the causes of disease; hypotheses about the treatment of disease; and hypotheses about how physical instruments can contribute to the diagnosis and treatment of disease. Generation of new hypotheses about health and disease often involves the creation of new concepts such as *virus*, *vitamin C*, and *X-ray*. I will now give examples of the different kinds of medical hypotheses and concepts.

Although medicine is largely concerned with the diagnosis, causes, and treatment of disease, a great deal of medical knowledge concerns the basic biological processes that support the healthy functioning of the body. The first reliable account of human anatomy was Vesalius's *On the Fabric of the Human Body*, published in 1543, which provided detailed illustrations of the structure of bones, muscles, organs, and blood vessels. His careful dissections produced discoveries about the structure of human bones that contradicted the accepted account of Galen, who had only dissected non-humans. The first major discovery in physiology was William Harvey's recognition in his 1628 book that blood circulates through the body as the result of the pumping action of the heart. Although cells were first observed in the seventeenth century, it took 200 years before the discovery and acceptance of the hypotheses that all living things are made of cells and that all cells arise from preexisting cells. During the twentieth century, many hypotheses about the functioning of the human body were generated and confirmed, establishing the fields of genetics and molecular biology that provided the basis for modern molecular understanding of the causes of health and disease. Table 11.1 summarizes some of the most important medical discoveries concerned with basic biological processes. All of these discoveries eventually contributed to the discovery of the causes and treatments of disease, with a delay of decades or even centuries. For example, van Leeuwenhoek's discovery of "little animals" such as bacteria only became medically important 200 years later with the development of the germ theory of disease. All of these basic medical discoveries involved hypotheses about biological structure or function, and some required the introduction of new concepts such as *cell*, *gene*, and *hormone*.

Table 11.1
Some major discoveries concerning medically important biological processes.

Decade	Discovery	Discoverer	Hypotheses
1540s	anatomy	Vesalius	bone structure, etc.
1620s	circulation	Harvey	blood circulates
1670s	bacteria	Leeuwenhoek	animalcules exist
1830s	cell theory	Schleiden, etc.	organs have cells
1860s	genetics	Mendel	inheritance
1900s	hormones	Bayliss, etc.	messaging
1950s	DNA	Watson, Crick	DNA structure
1950s	immune system	Lederberg, etc.	clonal deletion

Discoveries that are more specifically medical concern the causes of diseases. Until modern Western medicine emerged in the nineteenth century, the predominant world theories attributed disease to bodily imbalances, involving the humors of Hippocratic medicine, the *yin, yang,* and *chi* of traditional Chinese medicine, and the *doshas* of traditional Indian Ayurvedic medicine. Pasteur revolutionized the explanation of disease in the 1860s with the hypothesis that many diseases such as cholera are caused by bacteria. In the twentieth century, other diseases were connected with infectious agents, including viruses and prions. The nutritional causes of some diseases were identified in the early twentieth century, for example, how vitamin C deficiency produces scurvy. Auto-immune diseases require explanation in terms of malfunction of the body's immune system, as when multiple sclerosis arises from damage to myelin in the central nervous system. Some diseases such as cystic fibrosis have a simple genetic basis arising from inherited mutated genes, while in other diseases such as cancer the molecular/genetic causes are more complex. The general form of a hypothesis about disease causation is: disease D is caused by factor F, where F can be an external agent such as a microbe or an internal malfunction. Table 11.2 displays some of the most important discoveries about the causes of diseases.

The third kind of medical hypothesis, and potentially the most useful, concerns the treatment and prevention of disease. Hypotheses about treatment of disease based on traditional imbalance theories, for example, the use in Hippocratic medicine of bloodletting to balance humors, have been popular but unsubstantiated. In contrast, Edward Jenner's discovery in the

Table 11.2
Some major discoveries concerning the causes of diseases.

Decade	Discovery	Discoverer	Hypotheses
1840s	cholera	Snow	cholera is water-borne
1840s	antisepsis	Semmelweiss	contamination causes fever
1870s	germ theory	Pasteur, Koch	bacteria cause disease
1890s	tobacco disease	Ivanofsky, Beijerinck	viruses cause disease
1910s	cholesterol	Anichkov	cause of artherosclerosis
1960s	oncogenes	Varmus	cancer
1980s	prions	Prusiner	prions cause kuru
1980s	HIV	Gallo, Montagnier	HIV causes AIDS
1980s	*H. pylori*	Marshall, Warren	*H. pylori* causes ulcers

1790s that inoculation provides immunity to smallpox has saved millions of lives, as have the twentieth-century discoveries of drugs to counter the infectious properties of bacteria and viruses. The discovery of insulin in the 1920s provided an effective means of treating type one diabetes, which had previously been fatal. Treatments need not actually cure a disease to be useful: consider the contribution of steroids to diminishing the symptoms of autoimmune diseases, and the use of painkillers such as aspirin to treat various afflictions. Surgical treatments have often proved useful for treating heart disease and cancer.

It might seem that the most rational way for medicine to progress would be from basic biological understanding to knowledge of the causes of a disease to treatments for the disease. Often, however, effective treatments have been found long before deep understanding of the biological processes they affect. For example, aspirin was used as a painkiller for most of a century before its effect on prostaglandins was discovered, and antibiotics such as penicillin were in use for decades before it became known how they kill bacteria. Lithium provided a helpful treatment for bipolar (manic-depressive) disorder long before its mechanism of action on the brain was understood. On the other hand, some of the discoveries about causes listed in table 11.2 led quickly to therapeutic treatments, as when the theory that ulcers are caused by bacterial infection was immediately tested by treating ulcer patients with antibiotics (Thagard, 1999).

Table 11.3 lists some of the most important discoveries about medical treatments. These fall into several disparate subcategories, including prevention, surgical techniques, and drug treatments. Vaccination, antisepsis,

Table 11.3
Some major discoveries about treatments of diseases.

Decade	Discovery	Discoverer	Hypotheses
1790s	vaccination	Jenner	prevent smallpox
1840s	anesthesia	Long	reduce pain
1860s	antiseptic surgery	Lister	prevent infection
1890s	aspirin	Hoffman	treat pain
1890s	radiation treatment	Freund	remove cancer
1900s	Salvarsan	Ehrlich	cure syphilis
1900s	blood transfusion	Landsteiner	transfusion works
1920s	antibiotics	Fleming	mold kills bacteria
1920s	insulin	Banting	treat diabetes
1930s	sulfa drugs	Domagk	cure infection
1950s	birth control pill	Pincus, etc.	prevent pregnancy
1950s	transplants	Murray	kidney, lung
1950s	polio vaccination	Salk	prevent polio
1960s	IVF	Edwards	treat infertility
1980s	anti-retrovirals	various	slow HIV infection

and birth control pills serve to prevent unwanted conditions. Anesthesia, blood transfusions, organ transplants, and in vitro fertilization all involve the practice of surgery. Drug treatments include aspirin, antibiotics, and insulin. All of the treatments in table 11.3 are based on hypotheses about how an intervention can bring about improvements in a medical situation. A few involve new concepts, such as the concept *blood type*, which was essential for making blood transfusions medically feasible.

My fourth kind of medical discovery involves hypotheses about the usefulness of various instruments. I listed X-rays among the most important medical discoveries because of the enormous contribution that X-ray machines have made to the diagnosis of many ailments, from bone fractures to cancers. Other instruments of great medical importance are the stethoscope, invented in 1816, and techniques of testing blood for blood type, infection, and other medically relevant contents such as cholesterol levels. More recent instruments of medical significance include ultrasound scanners developed in the 1960s, computed tomography (CT) scanners invented in the 1970s, and magnetic resonance imaging (MRI) adopted in the 1980s. All of these instruments required invention of a physical device, which involved hypotheses about the potential usefulness of the device for identifying diseases and their causes. Table 11.4 lists some of the major

Table 11.4
Some major discoveries of diagnostic instruments.

Decade	Discovery	Discoverer	Hypotheses
1810s	stethoscope	Laennec	measure heart
1890s	X-rays	Roentgen	reveal bones
1900s	EKG	Einthoven	measure heart
1900s	tissue culture	Harrison	detect infections
1920s	cardiac catheterization	Forssman	inspect heart
1950s	radioimmunoassay	Yalow	analyze blood
1970s	CAT scans	Hounsfield	observe tissue
1970s	MRI scans	Lauterbur	observe tissue

medical discoveries involving physical instruments useful for the diagnosis of diseases. The origination of such instruments is perhaps better characterized as invention rather than discovery, but it still requires the generation of new hypotheses about the effectiveness of the instrument for identifying normal and diseased states of the body. For example, when Laennec invented the stethoscope, he did so because he hypothesized that a tube could help him better hear the operation of his patients' hearts.

The discovery of new hypotheses always requires the novel juxtaposition of concepts not previously connected. For example, the hypotheses that comprise the germ theory of disease connect a specific disease such as peptic ulcer with a specific kind of bacteria such as *Helicobacter pylori*. Construction of hypotheses requires the application and sometimes the generation of concepts. In the early stage of the bacterial theory of ulcers, Barry Marshall and Robin Warren associated the concepts *ulcer, cause,* and *bacteria,* and later their hypothesis was refined by specification of the bacteria via the concept of *H. pylori*. Other concepts of great importance in the history of discovery of the causes of disease include *vitamin, virus, autoimmune, gene,* and *oncogene*. Hence a theory of medical discovery will have to include an account of concept formation as well as an account of the generation of hypotheses. How this might work is discussed in the section below on psychological patterns.

All the medical discoveries so far discussed have involved the generation of specific new hypotheses. But there is another kind of more general medical breakthrough that might be counted as a discovery, namely, the development of new methods for investigating the causes and treatments of disease. Perhaps the first great methodological advance in the history of medicine was the Hippocratic move toward natural rather than magical or theological explanations of disease. The theory of humors was not, as it turned out millennia later, a very good account of the causes and treatments of disease, but at least it suggested how medicine could be viewed as akin to science rather than religion. In modern medicine, one of the great methodological advances was Koch's postulates for identifying the causes of infectious diseases (Brock, 1988, p. 180):

1. The parasitic organism must be shown to be constantly present in characteristic form and arrangement in the diseased tissue.

2. The organism which, from its behavior appears to be responsible for the disease, must be isolated and grown in pure culture.

3. The pure culture must be shown to induce the disease experimentally.

It turned out that these requirements, identified by Koch in the 1870s as part of his investigation of tuberculosis, are sometimes too stringent a requirement for inferring causes of infectious diseases, because some infectious agents are extremely difficult to culture and/or transmit. But the postulates have been useful for setting a high standard for identifying infectious agents. A third methodological breakthrough was the use, beginning only in the late 1940s, of controlled clinical trials in the investigation of the efficacy of medical treatments. Only decades later was it widely recognized that medical practices should ideally be determined by the results of randomized, double-blind, placebo-controlled trials, with the emergence of the movement for evidence-based medicine in the 1990s. None of these three methodological breakthroughs involves the discovery of particular medical hypotheses, but they all have been crucial to development of well-founded medical views about the causes and treatments of diseases.

Logical Patterns

Karl Popper published the English translation of his *Logik der Forschung* with the title *The Logic of Scientific Discovery*. The title is odd, for in the text he sharply distinguishes between the process of conceiving a new idea, and the methods and results of examining it logically (Popper, 1959, p. 21). The book is concerned with logic, not discovery. Like Reichenbach (1938) and many other philosophers of science influenced by formal logic, Popper thought philosophy should not concern itself with psychological processes of discovery. The term "logic" had come to mean "formal logic" in the tradition of Frege and Russell, in contrast to the broader earlier conception of logic as the science and art of reasoning. In John Stuart Mill's (1970) *System of Logic*, for example, logic is in part concerned with the mental processes of reasoning, which include inferences involved in scientific discovery.

If logic means just "formal deductive logic," then there is no logic of discovery. But N. R. Hanson (1958, 1965) argued for a broader conception of logic, which could be concerned not only with reasons for accepting a hypothesis but also with reasons for entertaining a hypothesis in the first place. He borrowed from Charles Peirce the idea of a kind of reasoning called "abduction" or "retroduction," which involves the introduction of hypotheses to explain puzzling facts. By abduction Peirce meant "the first starting of a hypothesis and the entertaining of it, whether as a simple interrogation or with any degree of confidence" (Peirce, 1931–1958, vol. 6, p. 358). Unfortunately, Peirce was never able to say what the first starting of a hypothesis amounted to, aside from speculating that people have an instinct for guessing right. In multiple publications, Hanson only managed to say that a logic of discovery would include a study of the inferential moves from the recognition of an anomaly to the determination of which types of hypothesis might serve to explain the anomaly (Hanson, 1965, p. 65). Researchers in artificial intelligence have attempted to use formal logic to model abductive reasoning, but Thagard and Shelley (1997) describe numerous representational and computational shortcomings of these approaches, such as that explanation is often not a deductive relation.

The closest we could get to a logical pattern of hypothesis generation for medical discovery, in the case of disease, would be something like:

Anomaly: People have disease D with symptoms S.

Hypothesis: Cause C can produce S.

Inference: So maybe C is the explanation of D.

For Pasteur, this would be something like:

Anomaly: People have cholera with symptoms of diarrhea, etc.

Hypothesis: Infection by a bacterium might cause such symptoms.

Inference: So maybe bacterial infection is the explanation of cholera.

Unfortunately, this pattern leaves unanswered the most interesting question about the discovery: how did Pasteur first come to think that infection by a bacterium might cause cholera? Answering this question requires seeing abduction as not merely a kind of deformed logic, but rather as a rich psychological process.

For Popper, Reichenbach, and even Hanson and Peirce, there is a sharp distinction between logic and psychology. This division is the result of the schism between philosophy and psychology that occurred because of the rejection by Frege and Husserl of psychologism in philosophy, as inimical to the objectivity of knowledge (see Thagard, 2000, ch. 1, for a historical review). Contemporary naturalistic epistemology in the tradition of Quine (1968) and Goldman (1986) rejects the expulsion of psychology from philosophical method. I will now try to show how richer patterns in medical discovery can be identified from the perspective of modern cognitive psychology.

Psychological Patterns

We saw in the last section that little can be said about discovery from the perspective of a philosophy of science that emphasizes logical structure and inference patterns. In contrast, a more naturalistic perspective that takes into account the psychological processes of practicing scientists has the theoretical resources to explain in much detail how discoveries come about. These resources derive from the development since the 1960s of the field of cognitive psychology, which studies the representations and procedures that enable people to accomplish a wide range of inferential tasks, from problem solving to language understanding. Starting in the 1980s, some philosophers of science have drawn on cognitive science to enrich accounts of the structure and growth of science knowledge (ch. 1). On this view, we should think of a scientific theory as a kind of mental representation that scientists can employ for many purposes such as explanation and discovery. Then scientific discovery is the generation of mental representations such as concepts and hypotheses.

I will not attempt a comprehensive account of all the cognitive processes relevant to discovery, nor will I attempt to apply them to explain the large number of discoveries listed in tables 11.1–11.4. Instead I will review a cognitive account of a single major medical discovery, the

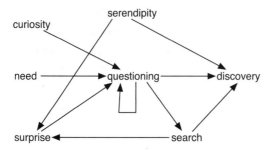

Figure 11.1
Psychological model of discovery. Adapted from Thagard (1999), p. 47.

realization by Barry Marshall and Robin Warren that most stomach ulcers
are caused by bacterial infection, for which they were awarded the 2005
Nobel Prize in Physiology or Medicine. Figure 11.1 depicts a general model
of scientific discovery developed as part of my account of the research of
Marshall and Warren (Thagard, 1999). Discovery results from two psychological processes, questioning and search, and from serendipity. Warren's
initial discovery of spiral gastric bacteria was entirely serendipitous, happening accidentally in the course of his everyday work as a pathologist.
Warren reacted to his observation of these bacteria with surprise, as it was
generally believed that bacteria could not long survive the acidic environment of the stomach. This surprise, along with general curiosity, led him
to generate questions concerning the nature and possible medical significance of the bacteria.

Warren enlisted a young gastroenterologist, Barry Marshall, to help him
search for answers about the nature and medical significance of the spiral
bacteria. After an extensive examination of the literature on bacteriology,
they concluded that the bacteria were members of a new species and genus,
eventually dubbed *Helicobacter pylori*. Here we see the origin of a new
concept: a mental representation of the bacteria that Warren observed
through a microscope. Marshall's questioning about the medical significance of these bacteria was driven not only by curiosity, but also by medical
needs, as he was aware that available medical treatments for stomach ulcers
using antacids were not very effective, diminishing symptoms but not
preventing recurrences.

Warren had observed that the bacteria were associated with inflamma-
tion of the stomach (gastritis), and Marshall knew that gastritis is associated

with peptic ulcer, so they naturally formed the hypothesis that the bacteria might be associated with ulcers. A 1982 study using endoscopy and biopsies found that patients with ulcers were far more likely to have *H. pylori* infections than patients without ulcers. They accordingly generated the hypothesis that the bacteria cause ulcers, by analogy with the many infectious diseases that had been identified since Pasteur. The natural psychological heuristic used here is something like: if A and B are associated, then A may cause B or vice versa. In order to show that A actually does cause B, it is desirable to manipulate A in a way that produces a change in B. Marshall and Warren were initially stymied, however, because of difficulties in carrying out the obvious experiments of giving animals *H. pylori* to induce ulcers and of giving people with ulcers antibiotics to try to kill the bacteria and cure the ulcers. Within a few years, however, they had discovered a regime involving multiple antibiotics that was effective at eradicating the bacteria, and by the early 1990s multiple international studies emerged that showed that such eradication often cured ulcers.

The discoveries of Marshall and Warren involve two main kinds of conceptual change. The first kind was introduction of the new concept of *Helicobacter pylori*, which was the result of both perceptual processes of observing the bacteria and of cognitive processes of conceptual combination. Originally they thought that the bacteria might belong to a known species, *Campylobacter*, hence the original name *Campylobacter pylori*, signifying that the new species inhabited the pylorus, the part of the stomach that connects to the duodenum. However, morphological and RNA analysis revealed that the new bacteria were very different from *Campylobacter*, so that they were reclassified as members of a new genus. Such reclassification is a second major kind of conceptual change, in that the discovery that bacteria cause ulcers produced a dramatic reclassification of the peptic ulcer disease. Previously, ulcers were viewed as metabolic diseases involving acid imbalance, or even, in older views as being psychosomatic diseases resulting from stress. Through the work of Marshall and Warren, peptic ulcers (except for some caused by nonsteroidal antiinflammatory drugs such as aspirin) were reclassified as infectious diseases, just like tuberculosis and cholera.

Thus the discovery of the bacterial theory of ulcers involved the generation and revision of mental representations. New concepts such as *H. pylori* were formed, and conceptual systems for bacteria and diseases were reorganized. Also generated were hypotheses, such as that bacteria cause ulcers and that ulcers can be treated with antibiotics. Both these sorts of representations can be produced by psychological processes of questioning, search, conceptual combination, and causal reasoning.

Analogy is a psychological process that often contributes to scientific discovery (Holyoak & Thagard, 1995). Marshall and Warren reasoned analogically when they thought that ulcers might be like more familiar infectious diseases. Other analogies have contributed to medical discoveries, such as Semmelweiss's mental leap from how a colleague became sick as the result of a cut during an autopsy to the hypothesis that childbed fever was being spread by medical students. Thagard (1999, ch. 9) describes other analogies that have contributed to medical discoveries, such as Pasteur's realization that disease is like fermentation in being caused by germs, and Funk's argument that scurvy is like beriberi in being caused by a vitamin deficiency. Thus analogy, like questioning, search, concept formation, and causal reasoning is an identifiable psychological pattern of discovery applicable to medical innovations.

Neural Patterns

The field of cognitive psychology is currently undergoing a major transformation in which the study of brain processes is becoming more and more central. Psychologists have long assumed that mental processing was fundamentally carried out by the brain, but the early decades of cognitive psychology operated independently of the study of the brain. This independence began to evaporate in the 1980s with the development of brain scanning technologies such as fMRI machines that enabled psychologists to observe brain activities in people performing cognitive tasks. Another major development in that decade was the development of connectionist computational models that used artificial neural networks to simulate psychological processes. (For a review of approaches to cognitive science, see Thagard, 2005a.) As illustrated by many journal articles and even the title of a recent textbook, *Cognitive Psychology: Mind and Brain* (Smith & Kosslyn, 2007), the field of cognitive science has become increasingly connected with neuroscience.

This development should eventually yield new understanding of scientific discovery. The psychological patterns of discovery described in the last section saw it as resulting from computational procedures operating on mental representations. From the perspective of cognitive neuroscience, representations are processes rather than things: they are patterns of activity in groups of neurons that fire as the result of inputs from other neurons. The procedures that operate on such mental representations are not much like the algorithms in traditional computer programs that inspired the early computational view of mind. Rather, if mental representations are patterns of neural activity, then procedures that operate on them are neural mechanisms that transform the firing activities of neural groups.

Accordingly, we ought to be able to generate new patterns of medical discovery construed in terms of neural activity. To my knowledge, the first neural network model of discovery is the highly distributed model of abductive inference described in chapter 3. Abninder Litt and I showed how to implement in a system of thousands of artificial neurons the simple pattern of inference from the occurrence of puzzling occurrence A and the knowledge that B can cause A to the hypothesis that B might have occurred. Representation of A, B, and *B causes A* is accomplished not by a simple expression or neuron, but by the firing activity of neural groups consisting of hundreds or thousand of neurons. The inference that B might have occurred is the result of systematic transformations of neural activity that take place in the whole system of neurons. This simple kind of abductive inference is not sufficient to model major medical discoveries, but it does appear appropriate for diagnostic reasoning of the following sort common in medical practice: This patient has ulcers; ulcers can be caused by bacterial infection; so maybe this patient has a bacterial infection. Much work remains to be done to figure out how neural systems can perform more complex kinds of inference, such as those that gave rise in the first place to the bacterial theory of ulcers.

On the neuroscience view of mental representation, a concept is a pattern of neural activity, so concept formation and reorganization are neural processes. In the development of the bacterial theory of ulcers, initial formation by Warren of the concept of spiral gastric bacteria seems to have been both perceptual and cognitive. The perceptual part began with the stimulation of Warren's retina by light rays reflected from his slides of stomach biopsies that revealed the presence of bacteria. At that point his perceptual representation of the bacteria was presumably a visual image constituted by neural activity in the brain's visual cortex. Warren's brain was able to integrate that visual representation with verbal representations consisting of other neural activities, thus linking the visual image to the verbal concepts *spiral, gastric,* and *bacteria.* But these concepts are not simply verbal, since they also involve representations that are partly visual, as is particularly obvious with the concept *spiral.* It is likely that for an experienced pathologist such as Warren the concepts *gastric* and *bacteria* are also partially visual: he had often seen pictures and diagrams of organs and microorganisms.

So how does the brain form concepts such as *spiral gastric bacteria of the kind observed through the microscope in Warren's samples*? I have previously described generation of new concepts as a kind of verbal conceptual combination, such as production of *sound wave* by combining the concepts of *sound* and *wave* (Thagard, 1988). But the neural process for Warren's new concept is considerably more complicated, as it requires integrating multiple representations including both verbal and nonverbal aspects. Here is a sketch of how this neural process might operate.

A crucial theoretical construct in cognitive psychology and neuroscience is *working memory* (Smith & Kosslyn, 2007; Fuster, 2002). Long-term memory in the brain consists of neurons and their synaptic connections. Working memory is a high level of activity in those groups of neurons that have been stimulated to fire more frequently by the current perceptual and inferential context that a person encounters. Then conceptual combination is the co-occurrence and coordination in working memory of a number of perceptual and verbal representations, each of which consists of patterns of neural activity. It is not yet well understood how this coordination occurs, but plausible hypotheses include neural synchronization (the patterns of neural activity become temporally related) and higher-level representations (patterns of neural activity in other neural groups represent patterns in the neural groups whose activity represents the original concepts). These two hypotheses may be compatible, since something like temporal coordination may contribute to the neural activity of the higher-order concept that ties everything together. Thus concept formation by perceptual-conceptual combination is a neural process involving the simultaneous activation and integration of previously unconnected patterns of neural activity (ch. 8).

This new account of multimodal conceptual combination goes well beyond the symbolic theory that I have applied to scientific discovery (Thagard, 1988). As Barsalou et al. (2003) argue, conceptual representations are often grounded in specific sensory modalities. For example, the concept *brown* is obviously connected with visual representation, as are more apparently verbal concepts like *automobile*, which may involve auditory and olfactory representations as well as visual ones. One advantage of theorizing at the neural level is that all of these kinds of verbal and sensory representations have the same underlying form: patterns of activity in neural groups. (See ch. 18 for more specifics.) Hence, newly generated concepts such as *brown automobile* and, more creatively, *gastric spiral bacteria*, can consist of neural activities that integrate verbal and sensory representations (see ch. 8).

Technological Patterns

My discussion of logical, psychological, and neural patterns of medical discovery has so far concerned the contributions of human beings to medical advances. But medical research increasingly relies on computers, not only to store information about biological systems but also to help generate new hypotheses about the causes and cures of disease. This section briefly sketches some emerging patterns of discovery that involve interactions between people and computers.

Computers have been essential contributors to projects involving basic biological processes, such as the Human Genome Project, completed in 2003. This project succeeded in identifying all the 20,000–25,000 genes in human DNA, determining the sequences of the 3 billion base pairs that make up human DNA, and storing the information in computer databases (Human Genome Project, 2006). All diseases have a genetic component, whether they are inherited or the result of an organism's response to its environment. Hence the information collected by the Human Genome Project should be of great importance to future investigations into the causes and treatments of a wide range of diseases. Such investigations would not be possible without the role of computers in sequencing, storing, and analyzing DNA information.

GenBank, the genetic sequence database compiled by the U.S. National Institutes of Health, contains over 50 million sequence records. These records include descriptions of many viruses, which proved useful in identifying the cause of the disease SARS that suddenly emerged in 2003. Within a few months, scientists were able to use the GenBank information and other technologies such as microarrays to determine that the virus responsible for SARS is a previously unidentified coronoavirus (Wang et al., 2003). Without computational methods for identifying the DNA structure of the virus associated with SARS and for comparing it with known structures, knowledge of the cause of SARS would have been greatly limited. Thus, computers are beginning to contribute to understanding of the causes of human diseases.

New technologies are also being developed to help find treatments for disease. Robots are increasingly used in automated drug discovery as part of the attempt to find effective new treatments, for example, new antibiotics that are not resistant to existing treatments. Lamb et al. (2006) describe their production of a "connectivity map," a computer-based reference collection of gene-expression profiles from cultured human cells treated with bioactive small molecules, along with pattern-matching software. This collection has the potential to reveal new connections among genes, diseases, and drug treatments. Thus, recent decades have seen the emergence of a new class of patterns of medical discovery in which human researchers cooperate with computers. Scientific cognition is increasingly distributed not only among different researchers but also among researchers and computers with which they interact (Thagard, 1999, 2006b; Giere, 2002). Because medical discovery is increasingly a matter of distributed cognition, the philosophy of medicine needs to investigate the epistemological implications of the collaborative, technological nature of medical research.

Conclusion

Although not much can be said about the *formal* logic of medical discovery, I hope to have shown that discovery is a live topic in the philosophy of medicine. We have seen that there are four kinds of discovery that require investigation, concerning basic biological processes, the causes of disease, the treatment of disease, and the development of new instruments for diagnosing and treating diseases. Psychological patterns of discovery include the development of new hypotheses by questioning, search, and causal reasoning, and the development of new concepts by combining old ones. Research in the burgeoning field of cognitive neuroscience is making it possible to raise, and begin to answer, questions about the neural processes that enable scientists to form hypotheses and generate concepts. In addition, philosophers can investigate how computers are increasingly contributing to new medical discoveries involving basic biological processes and the causes of disease. A major aim of the philosophy and cognitive science of medicine is to explain the growth of medical knowledge. Developing a rich, interdisciplinary account of the patterns of medical discovery should be a central part of that explanation.

IV Conceptual Change

According to the Google Books Ngram Viewer, the topic of conceptual change only became important during the 1960s. The rise in attention was the result of the radical claims made by Thomas Kuhn (1962) and Paul Feyerabend (1965), who suggested that the shifts between successive scientific theories are so radical that rational deliberation becomes impossible. Kuhn used the term "paradigm" in various ways, including for conceptual schemes that change dramatically. Kuhn identified major alterations in concepts such as *mass*, and dramatically pronounced that proponents of different paradigms live in different worlds.

Philosophers were aghast because this account of strong conceptual change seemed to erase common assumptions about scientific progress and rationality. In contrast, some sociologists were happy to see Kuhn shift attention toward the operations of scientific communities and away from philosophical norms. Kuhn himself resisted many of the more radical conclusions drawn from his work, and backed off from his most contentious claims about rationality (Kuhn, 1993). Nevertheless, *The Structure of Scientific Revolutions* was, as far as I can tell from Google Scholar, the most cited work in philosophy and in science studies during the twentieth century.

Fifty years after the publication of Kuhn's book, the topic of conceptual change remains important for several reasons. First, understanding conceptual development matters for the general appreciation of the structure and growth of scientific knowledge. As chapters 8 and 9 indicated, a major part of scientific development is the introduction of new concepts such as *gravity*, *molecule*, and *virus*. The historical record shows clearly that many new concepts are introduced as science develops, so we know that there is at least one important kind of conceptual change. It is also easy to identify concepts that have dropped out of scientific discourse as the theories that used them were superseded; abandoned concepts include *aether, phlo-giston,* and *spontaneous generation.*

Besides the addition and deletion of concepts, there are other kinds of conceptual change that the cognitive science of science can aim to understand. One of the most important is reclassification, where a concept is shifted from falling under one more general kind to falling under another. For example, one of the most important reclassifications in the history of science was a shift required by Darwin's theory of evolution by natural selection: counting humans as a kind of evolved animal rather than as a special class. Understanding the nature of these more radical kinds of conceptual change is crucial for addressing normative questions concerning rationality.

Second, conceptual development is not just an occurrence in the history of science, but also in the history of every child. Young children acquire thousands of new words per year, presumably along with the corresponding concepts. Hence at the very least they experience something like the addition of concepts that occurs in the history of science, although the major difference is that most of the new concepts are acquired from other people rather than generated anew as part of scientific discovery. Whether children experience other kinds of conceptual change remains an important topic for conceptual development. How close is the analogy between conceptual change in children and in the history of science? This question is still a live topic for discussion by philosophers and psychologists cited below.

The third reason why conceptual change is an important topic for the cognitive science of science results from its relevance to science education. As chapter 1 made clear, cognitive science has a normative as well as a descriptive dimension. The cognitive science of science should help not only to explain how children and other people acquire scientific concepts, but also to devise educational techniques that are more effective than current ones in enabling the acquisition of scientific knowledge. Educational psychologists have identified numerous shortfalls in science teaching. The cognitive science of science should be able to contribute to understanding why education is often hard, and also to developing more effective pedagogic techniques.

Facilitating conceptual change in this way is not just an issue for children and other students. For adults, scientific literacy is crucial for understanding and contributing to decisions about such pressing social issues as climate change, health care, and poverty. In a world of rapidly developing scientific ideas, scientific education is a process that needs to occur throughout a person's lifetime, not just during formal schooling. Hence the contributions that the cognitive science of science needs to make to understanding conceptual change should have broad social relevance.

Fortunately, the topic of conceptual change has received much attention in cognitive science, particularly from philosophers, developmental psychologists, and educational psychologists. Some philosophers have used the term "conceptual change" as almost synonymous with "belief revision," not paying much attention to the nature and development of concepts. In contrast, Nancy Nersessian (1984, 1992, 2008) has provided detailed investigations of scientific concepts that are very well informed by both the history of science and cognitive psychology. I have also tried to show the applicability of ideas from psychology and artificial intelligence to the understanding of conceptual change in the history of science (Thagard, 1992, 1999). My general conclusion with respect to the philosophical issues that generated concerns about conceptual change is this: Fine-grained and psychologically informed analysis of conceptual development does *not* support Kuhn's radical claims about scientific revolutions.

The leading developmental psychologist working on conceptual change has been Susan Carey (1985, 2009). But many other psychologists have done important experimental research on how children acquire new concepts, for example Elizabeth Spelke, Renee Baillargeon, and Susan Gellman. Issues about the extent to which children's concepts are innate rather than learned are still unresolved. Questions about the nature of conceptual change naturally depend on the general nature of concepts, a topic of great concern to cognitive psychologists (e.g., Murphy, 2002).

Cognitive and social psychologists have paid increasing attention in recent years to conceptual differences across cultures (Kitayama & Cohen, 2007), but they rarely address what might be involved in making the conceptual changes required to move from one culture to another. Such changes would also be a fitting topic for cognitive anthropology, which also, however, has been more concerned with conceptual differences than with conceptual changes (d'Andrade, 1995). Researchers in artificial intelligence have rarely addressed issues about conceptual change, but there is a large literature on computational methods for acquiring concepts from examples (e.g., Langley, 1996).

Conceptual change has been a major topic for psychologists concerned with science education (e.g., Chi, 2005, 2008; diSessa, 1988; Sinatra & Pintrich, 2003; Vosniadou, 2008; Vosniadou & Brewer, 1992). Most of their attention has been to physics and biology, which generate difficult questions about how to shift students from naïve, prescientific concepts to ones that are required by current scientific theories. It would be wonderful to have a general theory of conceptual change applicable to educational psychology, developmental psychology, and the history of science. But more investigation needs to be done to determine how much commonality exists among the mental structures and processes that support change in students, young children, and scientists.

In lieu of a treatise on conceptual change, the rest of Part IV provides four essays on the topic. Chapter 13 discusses the three most important concepts in biology, psychology, and medicine: *life*, *mind*, and *disease*. It shows how all three have changed from theological to qualitative to mechanistic understanding. Chapter 14 applies cognitive ideas about conceptual change to issues about science education concerning why it can be very difficult to inform and convince students and other people about the scientific merits of Darwin's theory of evolution. Chapter 15 provides a contrast between the conceptual systems of modern Western medicine and ancient Chinese medicine. Chapters 14 and 15 follow chapter 5 in observing that conceptual change has important emotional dimensions. Finally, conceptual change in medicine is discussed in chapter 16 with respect to ideas about mental illness, currently in transition from qualitative to mechanistic forms. In line with the mechanista approach advocated in chapter 1, this chapter makes it clear how conceptual change is about mechanisms as well as concepts.

This volume concludes in Part V with two newly written chapters that are also relevant to questions about conceptual change. Chapter 17 provides a novel way of thinking about scientific values and emotional conceptual change using a technique of cognitive-affective mapping that displays the emotional weightings of concepts as well as some of their cognitive relations. Finally, chapter 18 applies some of Chris Eliasmith's ideas from theoretical neuroscience to propose how scientific concepts such as *force*, *water*, and *cell* can be understood as neural processes.

13 Conceptual Change in the History of Science: Life, Mind, and Disease

Introduction

Biology is the study of life, psychology is the study of mind, and medicine is the investigation of the causes and treatments of disease. This chapter describes how the central concepts of life, mind, and disease have undergone fundamental changes in the past 150 years or so. There has been a progression from theological, to qualitative, to mechanistic explanations of the nature of life, mind, and disease. This progression has involved both theoretical change, as new theories with greater explanatory power replaced older ones, and emotional change, as the new theories brought reorientation of attitudes toward the nature of life, mind, and disease. After a brief comparison of theological, qualitative, and mechanistic explanations, I will describe how shifts from one kind of explanation to another have carried with them dramatic kinds of conceptual change in the key concepts in the life sciences. Three generalizations follow about the nature of conceptual change in the history of science: there has been a shift from conceptualizations in terms of simple properties to ones in terms of complex relations; conceptual change is theory change; and conceptual change is often emotional as well as cognitive.

The contention that historical development proceeds in three stages originated with the nineteenth-century French philosopher Auguste Comte, who claimed that human intellectual development progresses from a theological to a "metaphysical" stage to a "positive" (scientific) stage (Comte, 1970). The stages I have in mind are different from Comte's, so let me say what they involve. By the *theological* stage I mean systems of thought in which the primary explanatory entities are supernatural ones beyond the reach of science, such as gods, devils, angels, spirits, and souls.

For example, the concept of fire was initially theological, as in the Greek myth of Prometheus receiving fire from the gods. By the *qualitative* stage I mean systems of thought that do not invoke supernatural entities, but which postulate natural entities not far removed from what they are supposed to explain, such as vital force in biology. Early qualitative concepts of fire include Aristotle's view of fire as a substance and Epicurus's account of fire atoms. By the *mechanistic* stage I mean the kinds of developments now rapidly taking place in all of the life sciences in which explanations consist of identifying systems of interacting parts that produce observable changes. The modern concept of fire is mechanistic: combustion is rapid oxidation, the combination of molecules. Much more will be said about the nature of mechanistic, qualitative, and theological explanations in connection with each of the central concepts of life, disease, and mind. I will show how resistance to conceptual change derives both from (1) cognitive difficulties in grasping the superiority of mechanistic explanations to the other two kinds and (2) from emotional difficulties in accepting the personal implications of the mechanistic worldview. First, however, I want to review the general importance of the topic of conceptual change for the history and philosophy of science.

History and Philosophy of Science

Historians and philosophers of science are concerned to explain the development of scientific knowledge. On a naïve view, science develops by simple accumulation, piling fact upon fact. But this view is contradicted by the history of science, which has seen many popular theories eventually rejected as false, including: the crystalline spheres of ancient and medieval astronomy, the humoral theory of medicine, catastrophist geology, the phlogiston theory of chemistry, the caloric theory of heat, the vital force theory of physiology, the aether theories of electromagnetism and optics, and biological theories of spontaneous generation. Rejection of these theories has required abandonment of concepts such as *humor, phlogiston, caloric,* and *aether,* along with introduction of new theoretical concepts such as *germ, oxygen, thermodynamics,* and *photon.* Acceptance of a theory therefore often requires the acquisition and adoption of a novel conceptual system.

We can distinguish different degrees of conceptual change occurring in the history of science and medicine (Thagard, 1992, 1999, p. 150):

1. Adding a new instance of a concept, for example, a patient who has tuberculosis.

2. Adding a new weak rule, for example, that tuberculosis is common in prisons.

3. Adding a new strong rule that plays a frequent role in problem solving and explanation, for example, that people with tuberculosis have *Mycobacterium tuberculosis*.

4. Adding a new part-relation, for example, that diseased lungs contain tubercles.

5. Adding a new kind-relation, for example, differentiating between pulmonary and miliary tuberculosis.

6. Adding a new concept, for example, *tuberculosis* (which replaced the previous terms *phthisis* and *consumption*) or AIDS.

7. Collapsing part of a kind-hierarchy and abandoning a previous distinction, for example, realizing that phthisis and scrofula are the same disease, tuberculosis.

8. Reorganizing hierarchies by *branch jumping*, that is, shifting a concept from one branch of a hierarchical tree to another, for example, reclassifying tuberculosis as an infectious disease.

9. *Tree switching*, that is, changing the organizing principle of a hierarchical tree, for example, classifying diseases in terms of causal agents rather than symptoms.

The most radical kinds of conceptual change involve the last two kinds of major conceptual reorganization, as when Darwin reclassified humans as animals and changed the organizational principle of the tree of life to be evolutionary history rather than similarity of features.

Thus, understanding the historical development of the sciences requires attention to the different kinds of conceptual change that have taken place in the noncumulative growth of knowledge (see also Kuhn, 1962; Horwich, 1993; Laporte, 2004; Nersessian, 1992, 2008). I will now describe the central changes that have taken place in the concepts of life, mind, and disease.

Life

Theology

Theological explanations of life are found in the creation stories of many cultures, including the Judeo-Christian tradition's book of Genesis. According to this account God created grass, herbs, and fruit trees on the second day, swarms of birds and sea animals on the fifth day, and living creatures on land including humans on the sixth day. Other cultures worldwide have different accounts of how one or more deities brought the Earth and the living things on it into existence. These stories predate by centuries attempts to understand the world scientifically, which may only have begun with the thought of the Greek philosopher-scientist Thales around 600 BC. The stories do not attempt to tie theological explanations to details of observations of the nature of life. Thus, the first substage of the theological stage of the understanding of life is a matter of myth, a set of entertaining stories rather than a detailed exposition of the theological origins of life.

During the seventeenth and eighteenth centuries, there was a dramatic expansion of biological knowledge based on observation, ranging from the discovery by van Leeuwenhoek of microorganisms such as bacteria to the taxonomy by Carl Linnaeus of many different kinds of plants and animals. In the nineteenth century, attempts were made to integrate this burgeoning knowledge with theological understanding, including the compellingly written *Natural Theology* of William Paley (1963). Paley argued that, just as we explain the intricacies of a watch by the intelligence and activities of its maker, so we should explain the design of plants and animals by the actions of the creator. The eight volumes of the Bridgewater Treatises connected divine creation not only to the anatomy and physiology of living things, but also to astronomy, physics, geology, and chemistry. Nineteenth-century natural theology was a Christian enterprise, as theologians and believing scientists connected biological and other scientific observations in great detail with ideas drawn from the Bible. Unlike the purely mythical accounts found in many cultures, this natural-theology substage of theological explanations of life was tied to many facts about the biological world.

A third substage of theological understandings of life is the relatively recent doctrine of intelligent design that arose in the United States as a way of contesting Darwin's theory of evolution by natural selection without directly invoking Christian ideas about creation. Because the American constitution requires separation of church and state, public schools have not been allowed to teach Christian ideas about divine creation as a direct challenge to evolution. Hence in the 1990s there arose a kind of natural theology in disguise claiming to have a scientific alternative to evolution, the theory of intelligent design (see, e.g., Dembski, 1999). Its proponents claim that it is not committed to the biblical account of creation, but instead relies on facts about the complexity of life as pointing to its origins in intelligent causation rather than the mechanical operations of natural selection. American courts have, however, ruled that intelligent design is just a disguised attempt to smuggle natural theology into the schools.

Qualitative Explanations of Life

Unlike theological explanations, qualitative accounts do not invoke supernatural entities, but instead attempt to explain the world in terms of natural properties. For example, in the eighteenth century, heat and temperature were explained by the presence in objects of a qualitative element called caloric: the more caloric, the more heat. A mechanical theory of heat as motion of molecules only arose in the nineteenth century. Just as caloric was invoked as a substance to explain heat, qualitative explanations of life can be given by invoking a special kind of substance that inhabits living things. Aristotle, for example, believed that animals and plants have a principle of life (*psuche*) that initiates and guides reproductive, metabolic, growth, and other capacities (Grene & Depew, 2004).

In the nineteenth century, qualitative explanations of life became popular in the form of *vitalism*, according to which living things contain some distinctive force or fluid or spirit that makes them alive (Bechtel & Richardson, 1998). Scientists and philosophers such as Bichat, Magendie, Liebig, and Bergson postulated that there must be some sort of vital force that enables organisms to develop and maintain themselves. Vitalism developed as an opponent to the materialistic view, originating with the Greek atomists and developed by Descartes and his successors, that living things are like machines in that they can be explained purely in terms of the operation of their parts. Unlike natural theology, vitalism does not explicitly employ divine intervention in its explanation of life, but for vitalists such as Bergson there was no doubt that God was the origin of vital force.

Contrast the theological and vitalist explanation patterns.

Theological explanation pattern

Why does an organism have a given property that makes it alive?

Because God designed the organism to have that property.

Vitalist explanation pattern

Why does an organism have a given property that makes it alive?

Because the organism contains a vital force that gives it that property.

We can now examine a very different way of explaining life: in terms of mechanisms.

Mechanistic Explanations of Life

The mechanistic account of living things originated with Greek philosophers such as Epicurus, who wanted to explain all motion in terms of the interactions of atoms. Greek mechanism was limited, however, by the comparative simplicity of the machines available to them: levers, pulleys, screws, and so on. By the seventeenth century, however, more complicated machines were available, such as clocks, artificial fountains, and mills. In his 1664 *Treatise on Man*, Descartes used these as models for maintaining that animals and the bodies (but not the souls) of humans are nothing but machines explainable through the operations of their parts, analogous to the pipes and springs of fountains and clocks (Descartes, 1985). Descartes undoubtedly believed that living machines had been designed by God, but the explanation of their operations was in terms of their structure rather than their design or special vital properties. The pattern is something like this:

Mechanistic explanation pattern

Why does an organism have a given property that makes it alive?

Because the organism has parts that interact in ways that give it that property.

Normally, we understand how machines work because people have built them from identifiable parts connected to each other in observable ways.

In Descartes's day, mechanistic explanations were highly limited by lack of knowledge of the smaller and smaller parts that make up the body: cells were not understood until the nineteenth century. They were also limited by the simplicity of available machines to provide analogies to the complexities of biological organisms. By the nineteenth century, however, the cell doctrine and other biological advances made mechanistic explanations of life much more conceivable. But it was still utterly mysterious how different species of living things came to be, unless they were the direct result of divine creation. Various thinkers conjectured that species have evolved, but no one had a reasonable account of how they had evolved.

The intellectual situation changed dramatically in 1859, when Charles Darwin published *On the Origin of Species*. His great insight was not the concept of evolution, which had been proposed by others, but the concept of natural selection, which provided a mechanism that explained how evolution occurred. At first glance, natural selection does not sound much like a machine, but it qualifies as a mechanism because it consists of interacting parts producing regular changes. (For philosophical discussions of the nature of mechanisms, see Salmon, 1984; Bechtel & Richardson, 1993; Machamer, Darden & Craver, 2000; Bechtel & Abrahamsen, 2005.) The parts are individual organisms that interact with each other and with their environments. Darwin noticed that variations are introduced when organisms reproduce, and that the struggle for existence that results from scarcity of resources would tend to preserve those variations that gave organisms advantages in survival and reproduction. Hence variation plus the struggle for existence led to natural selection, which leads to the evolution of species. Over the past 150 years, the evidence for evolution by natural selection has accumulated to such an extent that it ought to be admitted that evolution is a fact as well as a theory.

Why then is there continuing opposition to Darwin's ideas? The answer is that the battle between evolution and creation is not just a competition between alternative theories of how different species came to be, but between different worldviews with very different emotional attachments. Theological views have limited explanatory power compared to science, but they have very strong emotional coherence because of their fit with people's personal goals, including comfort, immortality, morality, and social cohesion (Thagard, 2006a, ch. 14). People attach strong positive emotional valences to the key ingredients of creationist theories, including supernatural entities such as God and heaven. In contrast, evolution by natural selection strikes fundamentalist believers as atheistic and immoral.

Although Darwin conceived of a mechanism for evolution, he lacked a mechanistic understanding of key parts of it. In particular, he did not have a good account of how variations occurred and were passed on to offspring. Explanation of variation and inheritance required genetic theory, which (aside from Mendel's early ignored ideas) was not developed until the first part of the twentieth century. In turn, understanding of genetics developed in the second part of that century through discovery of how DNA provides a mechanism for inheritance. Today, biology is thoroughly mechanistic, as biochemistry explains how DNA and other molecules work, which explains how genes work, which explains how variation and inheritance work. The genomes of important organisms including humans have been mapped, and the burgeoning enterprise of proteomics is filling in the details of how genes produce proteins whose interactions explain all the operations required for the survival and reproduction of living things.

Hence what makes things alive is not a divine spark or vital force, but their construction out of organs, tissues, and individual cells that are alive. Cells are alive because their proteins and processes enable them to perform functions such as energy acquisition, division, motion, adhesion, signaling, and self-destruction. The molecular basis of each of these functions is increasingly well understood (Lodish et al., 2000). In turn, the behavior of molecules can be described in terms of quantum chemistry, which explains how the quantum-mechanical properties of atoms cause them to combine in biochemically useful ways. Thus, the development of biology over the past 150 years dramatically illustrates the shift from a theological to a qualitative to a mechanist concept of life. This shift has taken place because of an impressive sequence of mechanistic theories that provide deeper and deeper explanations of how living things work, from natural selection to genetics to molecular biology to quantum mechanics. This shift does not imply that there is only one fundamental level at which all explanation should take place: it would be pointless to try to give a quantum-mechanical explanation of why humans have large brains, as the quantum details are far removed from the historical environmental and biological conditions that produced the evolution of humans. It is enough, from the mechanistic point of view, that the lower-level mechanical operations are available in the background.

In sum, theoretical progress in biology has resulted from elaboration of progressively deeper mechanisms, while resistance to such progress results from emotional preferences for theological over mechanistic explanation. Similar resistance arises to understanding disease and mind mechanistically.

Disease

Theology

Medicine has both the theoretical goal of finding explanations of disease and the practical goal of finding treatments for them. As with conceptions of life, early conceptions of disease were heavily theological. Gods were thought to be sometimes the cause of disease, and they could be supplicated to provide relief from them. For example, in the biblical book of Exodus, God delivers a series of punishments, including boils, on the Egyptians for holding the Israelites captive. Hippocrates wrote around 400 BC challenging the view that epilepsy is a "sacred disease" resulting from divine action. Medieval Christians believed that the black plague was a punishment from God. In modern theology, diseases are rarely attributed directly to God, but there are still people who maintain that HIV/AIDS is a punishment for homosexuality. But even if most people now accept medical explanations of the causes of disease, there are many who pray for divine intervention to help cure the maladies of people they care about. Hence in religious circles the concept of disease remains at least in part theological.

Qualitative Explanations of Disease

The ancient Greeks developed a naturalistic account of diseases that dominated Western medicine until the nineteenth century (Hippocrates, 1988). According to the Hippocratics, the body contains four humors: blood, phlegm, yellow bile, and black bile. Health depends on having these humors in correct proportion to each other. Too much bile can produce various fevers, and too much phlegm can cause heart or brain problems. Accordingly, diseases can be treated by changing the balance of humors, for example, by opening the veins to let blood out.

Traditional Chinese medicine, which is at least as ancient as the Hippocratic approach, is also a balance theory, but with *yin* and *yang* instead of the four humors. On the Chinese view, yin and yang are the two opposite but complementary forces that constitute the entire universe (see chapter 15). Diseases arise when there is an imbalance of *yin* and *yang* inside the body. Treatments such as herbs can restore the balance of *yin* and *yang*. Whereas the Hippocratic tradition used extreme physical methods such as blood-letting, emetics, and purgatives to restore the balance of the four humors, traditional Chinese medicine uses relatively benign herbal treatments to restore the balance of *yin* and *yang*. Unlike Hippocratic medicine, which has been totally supplanted by Western scientific approaches, traditional Chinese medicine is still practiced in China and is often favored by Westerners looking for alternative medical treatments.

Similarly, traditional Indian Ayurvedic medicine has attracted a modern following through the writings of gurus such as Deepak Chopra. On this view, all bodily processes are governed by three main *doshas*: *vata* (composed of air and space), *pitta* (composed of fire and water), and *kapha* (composed of earth and water). Too much or too little of these elements can lead to diseases, which can be treated by diet and exercise. There is no empirical evidence for the existence of the *doshas* or for their role in disease, but people eagerly latch onto Chopra's theories for their promise that good health and long life can be attained merely by making the right choices. Just as creationism survives because it fits with peoples personal motivations, so traditional Chinese and Ayurvedic theories survive because they offer appealing solutions to scary medical problems.

The three balance theories described in this section are clearly not theological, because they do not invoke divine intervention. But they are also not mechanical, because they do not explain the causes of diseases in terms of the regular interaction of constitutive parts. They leave utterly mysterious how the interactions of humors, *doshas*, or *yin* and *yang* can make people sick. In contrast, modern Western medicine based on contemporary biology provides mechanistic explanations of a very wide range of diseases.

Mechanistic Explanations of Disease

Modern medicine began in the 1860s, when Pasteur and others developed the germ theory of disease. Bacteria had been observed microscopically in the 1670s, but their role in causing diseases was not suspected until Pasteur realized that bacteria are responsible for silkworm diseases. Bacteria were quickly found to be responsible for many human diseases, including cholera, tuberculosis, and gonorrhea. Viruses were not observed until the invention of the electron microscope in 1939, but are now known to be the cause of many human diseases such as influenza and measles (Thagard, 1999).

The germ theory of disease provides mechanistic explanations in which bacteria and viruses are entities that interact with bodily parts such as organs and cells that are infected. Unlike vague notions like *yin*, *yang*, and *doshas*, these entities can be observed using microscopes, as can their presence in bodily tissues. Thus an infected organism is like a machine that has multiple interacting parts. The germ theory of disease is not only theoretically useful in explaining how many diseases arise, it is also practically useful in that antimicrobial drugs such as penicillin can cure some diseases by killing the agents that cause them.

As we saw for biological explanations, it is a powerful feature of mechanistic explanations that they decompose into further layers of mechanistic explanations. Pasteur had no idea how bacteria manage to infect organs, but molecular biology has in recent decades provided detailed accounts of how microbes function. For example, when the new disease SARS was identified in 2003, it took only a few months to identify the coronavirus that causes it and to sequence the virus's genes that enable it to attach themselves to cells, infect them, and reproduce. In turn, biochemistry explains how genes produce the proteins that carry out these functions. Thus the explanations provided by the germ theory have progressively deepened over the almost one and half centuries since it was first proposed. Chapter 6 argues that this kind of ongoing deepening is a reliable sign of the truth of a scientific theory.

Not all diseases are caused by germs, but other major kinds have been amenable to mechanistic explanation. Nutritional diseases such as scurvy are caused by deprivation of vitamins, and the mechanisms by which vitamins work are now understood. For example, vitamin C is crucial for collagen synthesis and the metabolism and synthesis of various chemical structures, which explains why its deficiency produces the symptoms of scurvy. Some diseases are caused by the immune system becoming overactive and attacking parts of the body, as when white blood cells remove myelin from axons between neurons, producing the symptoms of multiple sclerosis. Other diseases such as cystic fibrosis are directly caused by genetic factors, and the connection between mutated genes and defective metabolism is increasingly well understood. The final major category of human disease is cancer, and the genetic mutations that convert a normal cell into an invasive carcinoma, as well as the biochemical pathways that are thereby affected, are becoming well mapped out (Thagard, 2003, 2006b).

Despite the progressively deepening mechanistic explanation of infectious, nutritional, autoimmune, and genetic diseases, there is still much popular support for alternative theories and treatments such as traditional Chinese and Ayurvedic medicine. The reasons for the resistance to changes in the concept of disease from qualitative to mechanistic are both cognitive and emotional. On the cognitive side, most people simply do not know enough biology to understand how germs work, how vitamins work, how the immune system works, and so on. Hence much simpler accounts of imbalances among a few bodily elements are appealing. On the emotional side, there is the regrettable fact that modern medicine still lacks treatment for many human diseases, even ones like cancer whose biological mechanisms are quite well understood. Alternative disease theories and therapies offer hope of inexpensive and noninvasive treatments. For example, naturopaths attribute diseases to environmental toxins that can be cleared by diet and other simple therapies, providing people with reassuring explanations and expectations about their medical situation. Hence resistance to conceptual change about disease, like resistance concerning life, is often as much emotional as cognitive. The same is true for the concept of mind.

Mind

Theology
For the billions of people who espouse Christianity, Islam, Hinduism, and Buddhism, a person is much more than a biological mechanism. According to the book of Genesis, God formed man from the dust of the ground and breathed into his nostrils, making him a living soul. Unlike human bodies, which rarely last more than 100 years, souls have the great advantage of being indestructible, which makes possible immortality and (according to some religions) reincarnation. Because most people living today believe that their souls will survive the demise of their bodies, they have a concept of a person that is inherently dualistic, according to which people consist of both a material body and a spiritual soul.

We saw that Descartes argued that bodies are machines, but he maintained that minds are not mechanically explainable. His main argument for this position was a thought experiment: he found it easy to imagine himself without a body, but impossible to imagine himself not thinking

(Descartes, 1985). Hence he concluded that he was essentially a thinking being rather than a bodily machine, thereby providing a conceptual argument for the theological view of persons as consisting of two distinct substances, with the soul being much more important than the body. Descartes thought that the body and soul were able to influence each other through interaction in the brain's pineal gland.

The psychological theories of ordinary people are thoroughly dualist, assuming that consciousness and other mental operations belong fundamentally to the soul rather than the brain. Legal and other institutions assume that people inherently have the capacity for free will, which applies to actions of the soul rather than to processes occurring in the brain through interaction with other parts of the body and the external environment. Such freedom is viewed as integral to morality, making it legitimate to praise or blame people for their actions.

Notice how tightly the theological view of the mind as soul fits with the biological theory of creation. Life has theological rather than natural origins, and God is also responsible for a special kind of life: humans with souls as well as bodies. Gods and souls are equally supernatural entities.

Qualitative Explanations of Mind

Postulating souls with free will does not enable us to say much about mental operations, and many thinkers have used introspection (self-observation) to describe the qualitative properties of thinking. The British empiricist philosophers Locke and Hume claimed that minds function by the associations of ideas that are ultimately derived from sense experience. When Wilhelm Wundt originated experimental psychology in the 1870s, his observational method was still primarily introspective, but was much more systematic and tied to experimental interventions than ordinary self-observation.

Many philosophers have resisted the attempt to make the study of mind scientific, hoping that a purely conceptual approach could help us to understand thinking. Husserl founded phenomenology, an a priori attempt to identify essential features of thought and action. Linguistic philosophers such as J. L. Austin thought that attention to the ordinary uses of words could tell us something about the nature of mind. Analytic philosophers have examined everyday mental concepts such as belief and desire, under the assumption that people's actions are adequately explained as the result of people's beliefs and desires. Thought experiments survive as a popular philosophical tool for determining the essential features of thinking, for example, when Chalmers (1996) uses them to argue for a nontheological version of dualism in which consciousness is a fundamental part of the universe like space and time.

Thought experiments can be helpful for generating hypotheses that suggest experiments, but by themselves they provide no reason to believe those hypotheses. For every thought experiment there is an equal and opposite thought experiment, so the philosophical game of imagining what might be the case tells us little about the nature of minds and thinking. Introspective, conceptual approaches to psychology are appealing because they are much less constrained than experimental approaches and do not require large amounts of personnel and apparatus. They generate no annoying data to get in the way of one's favorite prejudices about the nature of mind. However, they are very limited in how much they can explain about the capacities and performance of the mind. Fortunately, mechanistic explanations based on experiments provide a powerful alternative methodology.

Mechanistic Explanations of Mind

Descartes thought that springs and other simple mechanisms suffice to explain the operation of bodies, but he drew back from considering thinking mechanistically. Until the second half of the twentieth century, these mechanical models of thinking such as hydraulic fluids and telephone switchboards seemed much too crude to explain the richness and complexity of human mental operations. The advent of the digital computer provided a dramatic innovation in ways of thinking about the mind. Computers are obviously mechanisms, but they have unprecedented capacities to represent and process information. In 1956, Newell, Shaw, and Simon (1958) developed the first computational model of human problem solving. For decades, the computer has provided a source of analogies to help understand many aspects of human thinking, including perception, learning, memory, and inference (Thagard, 2005a). On the computational view of mind, thinking occurs when algorithmic processes are applied to mental representations that are akin to the data structures found in the software that determines the actions of computer hardware.

However, as von Neumann (1958) noted early on, digital computers are very different from human brains. They nevertheless have proved useful for developing models of how brains work, ever since the 1950s. But in the 1980s there was an upsurge of development of models of brain-style computing, using parallel processing among simple processing elements roughly analogous to neurons (Rumelhart & McClelland, 1986). Churchland and Sejnowski (1992) and others have argued that neural mechanisms are computational, although of a rather different sort than those found in digital computers. More biologically realistic, computational models of neural processes are currently being developed (e.g., Eliasmith & Anderson, 2003). Efforts are increasingly made to relate high-level mental operations such as rule-based inference to neural structures and processes (e.g., Anderson et al., 2004; J. R. Anderson, 2007; Eliasmith, forthcoming). Thus neuroscience, along with computational ideas inspired by neural processes, provides powerful mechanistic accounts of human thinking.

Central to modern cognitive science is the concept of *representation*, which has undergone major historical changes. From a theological perspective, representations such as concepts and propositions are properties of spiritual beings, and thus are themselves nonmaterial objects. Modern cognitive psychology reclassifies representations as material things, akin to the data structures found in computer programs. Most radically, cognitive neuroscience reclassifies representations as *processes*, namely, patterns of activity in neural networks in the brain. Thus the history of cognitive science has required *branch jumping*, which I earlier listed as one of the most radical kinds of conceptual change. It is too soon to say whether cognitive neuroscience will also require *tree switching*, a fundamental change in the organizing principles by which mental representations are classified.

We saw in discussing life and disease how mechanistic explanations are decomposable into underlying mechanisms. At the cognitive level, we can view thinking in terms of computational processes applied to mental representations, but it has become possible to deepen this view by considering neurocomputational processes applied to neural representations. In turn, neural processes—the behavior of neurons interacting with each other—can be explained in terms of biochemical processes. The study of mind, like the study of life and disease, is increasingly becoming molecular

(Thagard, 2003, 2006a, ch. 7). That does not mean that the only useful explanations of human thinking will be found at the molecular level, because various phenomena are more likely to be captured by mechanisms operating at different levels. For example, rule-based problem solving may be best explained at the cognitive level in terms of mental representations and computational procedures, even if these representations and procedures ultimately derive from neural and molecular processes.

Indeed, a full understanding of human thinking needs to consider higher as well as lower levels. Many kinds of human thinking occur in social contexts, involving social mechanisms such as communication and other kinds of interaction. Far from it being the case that the social reduces to the cognitive which reduces to the neural which reduces to the molecular, sometimes what happens at the molecular level needs to be explained by what happens socially. For example, a social interaction between two people may produce very different kinds of neurotransmitter activity in their brains depending on whether they like or fear each other.

Of course, there is a great deal about human thinking that current psychology and neuroscience cannot yet explain. Although perception, memory, learning, and inference are increasingly subject to neurocomputational explanation, puzzles such as consciousness remain, where there are only sketches of mechanisms that might possibly be relevant. Such sketchiness gives hope to those who are opposed for various religious or ideological reasons to the provision of mechanistic explanations of the full range of human thought. From a theological perspective that assumes the existence of souls, full mechanistic explanation of thinking is impossible as well as undesirable. The undesirability stems from the many attractive features of supernatural souls, particularly their immortality and autonomy. Adopting a mechanistic view of mind requires abandoning or at lease modifying traditional ideas about free will, moral responsibility, and eternal rewards and punishment. This threat explains why the last fifty years of demonstrable progress in mechanistic, neurocomputational explanations of many aspects of thought are ignored by critics who want to maintain traditional attitudes. Change in the concept of mind, as with life and disease, is affected not only by cognitive processes such as theory evaluation, but also by emotional processes such as motivated inference. In the next section I will draw some more general lessons about conceptual change in relation to science education.

Conceptual Change

Of course, many other important concepts occur in the history of science besides life, mind, and disease, and there is much more to be said about other kinds of conceptual change (see, e.g., Thagard, 1992). But, because the concepts of life, mind, and disease are central, respectively, to biology, psychology, and medicine, they provide a good basis for making some generalizations about conceptual change in the history of science that can be tested against additional historical episodes. The commonalities in ways in which these three concepts have developed are well worth noting.

In all cases, there has been a shift from conceptualizations in terms of simple properties to ones in terms of complex relations. Prescientifically, life could be viewed as a special property that distinguished living from nonliving things. This property could be explained in terms of divine creation or some vital force. In contrast, the mechanistic view of biology considers life as a whole complex of dynamic relations, such as the metabolism and reproduction of cells. Life is no one thing, but rather the result of many different mechanical processes. Similarly, disease is not a simple problem that can be explained by divine affliction or humoral imbalance, but rather is the result of many different kinds of biological and environmental processes. Diseases have many different kinds of causes—microbial, genetic, nutritional, and autoimmune, each of which depends on many underlying biological mechanisms. Even more strikingly, mind is not a simple thing, a noncorporeal soul, but rather the result of many interacting neural structures and processes. Thus, the conceptual developments of biology, psychology, and medicine have all required shifts from thinking of things in terms of simple properties to thinking of them in terms of complexes of relations. Students who encounter scientific versions of their familiar everyday concepts of life, mind, and disease need to undergo the same kind of shift. Chi (2005) describes the difficulties that arise for students in understanding emergent mechanisms, ones in which regularities arise from complex interactions of many entities. Life, mind, and disease are all emergent processes in this sense and therefore subject to the difficult learning challenges that Chi reports in other domains.

The shift in understanding life, mind, and disease as complex mechanical relations rather than as simple substances or properties is an example of what I earlier called branch jumping, reclassification by shifting a concept from one branch of a hierarchical tree to another. The tree here is ontological, a classification of the fundamental things thought to be part of existence. Life, for example, is no longer a kind of special property, but rather a kind of mechanical process. Mind is another kind of mechanical process, not a special substance created by God. Many more mundane cases of branch jumping have occurred as the life sciences develop, for example, the reclassification in the 1980s of peptic ulcers as infectious diseases (Thagard, 1999).

Most radically, the shift from theological to qualitative to mechanistic conceptions of life, mind, and disease also involved tree switching, changing the organizing principle of a hierarchical tree. From a mechanistic perspective, we classify things in terms of their underlying parts and interactions. Darwin's mechanism of evolution by natural selection yielded a whole new way of classifying species, by historical descent rather than similarity. Later, the development of molecular genetics provided another new way of classifying species in terms of genetic similarity. Similarly, diseases are now classified in terms of their causal mechanisms rather than surface similarity of symptoms, for example, as infectious or autoimmune diseases. More slowly, mental phenomena such as memory are becoming classified in terms of underlying causal mechanisms such as different kinds of neural learning (Smith & Kosslyn, 2007). Thus, conceptual change in the life sciences has involved both branch jumping and tree switching.

Another important general lesson we can draw from the development of concepts of life, mind, and disease is that conceptual change in the history of science is theory change. Scientific concepts are embedded in theories, and it is only by the development of explanatory theories with broad empirical support that it becomes reasonable and in fact intellectually mandatory to adopt new complexes of concepts. The current scientific view of life depends on evolutionary, genetic, and molecular theories, just as the current medical view of disease depends on molecular, microbial, nutritional, and other well-supported theories. Similarly, our concept of mind should be under constant revision as knowledge accumulates about the neurocomputational mechanisms of perception, memory, learning, and inference. In all these cases, it would have been folly to attempt to begin investigation with a precise definition of key concepts, because what matters is the development of explanatory theories rather than conceptual neatness. After some theoretical order has been achieved, it may be possible to tidy up a scientific field with some approximate definitions. But if theoretical advances have involved showing that phenomena are much more complicated than anyone suspected, and that what were thought to be simple properties are in fact complexes of mechanical relations, then definitions are as pointless at later stages of investigation as they are distracting at early stages.

My final lesson about conceptual change in the history of science is that, especially in the sciences most deeply relevant to human lives, conceptual change is emotional as well as cognitive. The continuing resistance to mechanistic explanations of life, mind, and disease is inexplicable on purely cognitive grounds, given the enormous amount of evidence that has accumulated for theories such as evolution by natural selection, the germ theory of disease, and neurocomputational accounts of thinking. Although the scientific communities have largely made the emotional shifts necessary to allow concepts and theories to fit with empirical results, members of the general population, including many science students, have strong affective preferences for obsolete theories such as divine creation, alternative medicine, and soul-based psychology. Popular concepts of life, mind, and disease are tightly intertwined: God created both life and mind and can be called on to alleviate disease. Hence conceptual change can require not just rejection of a single theory in biology, psychology, and medicine, but rather replacement of a theological worldview by a scientific, mechanist one. For many people, such replacement is horrific, because of the powerful emotional appeal of the God-soul-prayer conceptual framework. Hence the kind of theory replacement required to bring about conceptual change in biology, psychology, and medicine is not just a matter of explanatory coherence, but requires changes in emotional coherence as well (for a theory of emotional coherence, see Thagard, 2000, 2006a).

From this perspective, science education inevitably involves cultural remediation and even psychotherapy, in addition to more cognitive kinds of instruction. The transition from theological to qualitative to mechanistic explanations of phenomena is cognitively and emotional difficult, but crucial for scientific progress, as we have seen for the central concepts of life, mind, and disease.

14 Getting to Darwin: Obstacles to Accepting Evolution by Natural Selection

Paul Thagard and Scott Findlay

Introduction

Darwin's theory of evolution by natural selection is widely recognized by scientists as the cornerstone of modern biology, but many people, especially in the United States, remain skeptical (Miller, Scott & Okamoto, 2006). According to a recent poll in Canada's oil-rich province of Alberta, more people there believe that humans were created by God within the last 10,000 years than believe that humans evolved from less advanced life forms over millions of years (Breakenridge, 2008). Science educators in several countries have reported serious problems in enabling students to understand and accept evolution by natural selection (e.g., R. D. Anderson, 2007; Blackwell, Powell & Dukes, 2003; Deniz, Donnelly & Yilmaz, 2008; Evans, 2008; Hokayem & BouJaoude, 2008; Sinatra et al., 2003).

At first glance, adoption of Darwin's theory should be straightforward. First, people can be instructed in basic ideas about how genetic variation and competition among individual organisms produce natural selection for characteristics that increase survival and reproduction in particular environments. Second, they can be led to see how natural selection has produced many kinds of evolutionary change, from the development of well-functioning organs to the occurrence in bacteria of resistance to antibiotics. Third, from presentation of the many kinds of evidence that have been explained using evolution by natural selection, people should be able to appreciate that Darwin's theory has far greater explanatory power than any available alternatives and therefore should be accepted.

Unfortunately, for science students and the population in general, there are major psychological obstacles to accepting Darwin's theory. Cognitive obstacles to adopting evolution by natural selection include conceptual difficulties, methodological issues, and coherence problems that derive from the intuitiveness of alternative theories. The main emotional obstacles to accepting evolution are its apparent conflict with valued beliefs about God, souls, and morality. Using a theory of cognitive and emotional belief revision, we will make suggestions about what can be done to improve acceptance of Darwinian ideas.

Cognitive Obstacles

Darwin's theory of evolution by natural selection should be accepted by everyone because it provides a far better explanation of a wide range of biological phenomena than available alternatives. Resistance to the theory is partly the result of emotional factors that we will discuss later. But first we will review cognitive factors that make it difficult for many people to understand and accept Darwin. Understanding his theory requires grasping its central concepts such as natural selection and appreciating how evolution by natural selection can explain many biological phenomena such as speciation. Accepting the theory (in philosophical terminology) would be the further step of judging it to be true.

Conceptual Difficulties

At first glance, the concepts required to understand evolution by natural selection should not be particularly hard to grasp. Such concepts as evolutionary change, genetic variation, struggle for existence, and natural selection can all be described qualitatively, without the mathematical complexities that impede understanding of major theories in physics such as relativity and quantum mechanics. But Darwin's theory concerns biological processes that are statistical and emergent in ways that do not fit well with commonsense explanations.

The two kinds of explanations that people are most familiar with are intentional ones and simple mechanical ones. Intentional explanations are familiar from social experience, for example, when individuals' decisions about careers are explained in terms of their beliefs and desires. Everyday mechanical explanations depend on causal relations that can be captured by straightforward if-then rules, such as if a bicycle's pedals turn, then its crank, chain, and wheels also turn. Thus the operation and occasional breakdowns of bicycles are amenable to simple explanations in terms of the parts and interactions of its pedals, crank, chain, and wheels.

In contrast, evolutionary explanations are inherently statistical, involving probabilistic changes in genes and species. They require what the eminent evolutionary biologist Ernst Mayr (1982) called "population thinking," which he contrasted with more natural typological thinking in terms of essences. Such statistical thinking is not natural to human cognition, as probability theory and the theory of statistical inference have only been invented in recent centuries (Hacking, 1975). Few students in high schools and universities are exposed to statistical thinking, so it is not surprising that people have difficulty thinking of biological species in statistical terms. Sophisticated probabilistic thinking in science only developed in the late nineteenth century in the field of statistical mechanics, which inspired twentieth-century work on population genetics. Psychological research has shown that conceptualization of species in terms of essences (inner natures) is an impediment to understanding natural selection (Shtulman, 2006; Shtulman & Schulz, 2008). Even when students accept evolution, they may continue to believe that it operates on a species' essence rather than on its members.

In addition to being inherently statistical, evolutionary explanations require appreciation of emergent processes, ones in which large effects result from small operations that are qualitatively different. In a simple mechanism such as a bicycle, the motion of the whole system can easily be seen to result from the motions of its parts. An emergent property is one that belongs to a system but not to any of the system's parts (Bunge, 2003). Such properties are ubiquitous in the natural world, at many levels, including molecules, cells, organs, organisms, and societies. To take a very simple example, molecules of sodium chloride (table salt) have properties such as stimulation of taste buds that are different from and hard to predict from the properties of the sodium and chlorine atoms that constitute them. An emergent process is one that produces emergent properties. The evolution of a new species is an emergent process arising from small genetics changes in a large number of individuals until eventually there exists a population of organisms no longer capable of interbreeding with members of previous species. Students need to understand how natural selection can lead not only to gradual changes within a population, but also to speciation—the emergence of a qualitatively different kind of organism.

Chi (2005, 2008) has reviewed the learning difficulties encountered by students who need to understand emergent processes in physics. Students naturally understand entities and their properties, such as animals that have sizes and colors. More difficult is appreciating processes that occur over time, but these are not so difficult as long as they are direct, where each change has an identifiable causal agent. In emergent processes, however, changes come about because of many small interactions, for example, the diffusion of oxygen from the lungs to the blood vessels, or migrating geese flying in a V-formation. Students' failure to understand how species can emerge without direction is part of widespread inability to understand emergent processes. Grasping speciation as the result of evolution by natural selection is made even more difficult by the fact that it occurs over a much larger time scale—usually thousands or millions of years—than students have previously encountered. Because of the relative security of modern cultures, the evolutionary pressures that produced human speciation are no longer evident.

Genetic changes are slow and statistical, in contrast to simple mechanical or intentional changes that are fast and direct. Students are naturally drawn to Lamarckian explanations in terms of the heritability of acquired characteristics, which is similar to the rapid, directed changes that take place in people's personal lives and cultural developments. Darwin lacked a good theory of how variation in populations can occur and be transmitted to future generations, a gap filled in the twentieth century by genetics. But most students are not well equipped to appreciate the nondirected nature of genetic mutation or the statistical nature of population genetics. Students also have a natural inclination toward purpose-based teleological explanations akin to familiar intentional explanations (Kampourakis & Zogza, 2008).

Historically and currently, one of the greatest obstacles to the acceptance of evolution is the claim that human thought is a product of it. Alfred Wallace, who discovered natural selection independently of Darwin, was never able to accept that it applied to minds, which he thought had an irreducible spirituality. Students today find the most implausible aspect of Darwin's theory to be the suggestion that it could provide a way of accounting for the operations of human minds (Ranney & Thanukos,

2010). Here students have a double difficulty: not only is evolution an emergent process on the Darwinian account, but thinking is also an emergent process on the account currently being developed in neuroscience. Thoughts are the result of the interactions of billions of neurons, each of which fires as the result of the interactions of thousands of genes, proteins, and other molecules. To believe that human thinking evolved by natural selection, a student has to be able to grasp several levels of emergent processes, including:

1. how evolution by natural selection could have produced the human brain, and

2. how interactions among neurons and neurochemicals could produce human thought.

Thus, the human mind is an emergent process resulting from an emergent process! So it is small wonder that students and lay persons, not to mention many contemporary philosophers, have great difficulty imagining how mind could be the result of brain structures arising from natural selection.

Contrary to commonsense psychology, we know from cognitive psychology and neuroscience that thinking is largely an unconscious process. People may think they understand their own minds through the conscious process of introspection, but cognitive research has shown that self-observation provides a very limited and often misleading source of information about how the mind works. In particular, conscious experience tells us little about the evolved biological processes that make consciousness possible. Thagard and Aubie (2008) offer a neurocomputational model of emotional consciousness that explains how many interacting brain areas can generate such emotions as happiness. But many people find it hard to conceive that conscious experience can be the result of biological processes.

Methodological Difficulties

Even if students manage to understand the complex process of evolution by natural selection, the question still arises how or why they should come to believe it. Within philosophy of science, there are contradictory accounts of what would be involved in accepting Darwin's theory. Here are just four:

1. A follower of the philosopher of science Karl Popper (1959) could say that students should not accept Darwin's theory at all, for two reasons.

First, no one should ever accept a scientific theory, since scientific method consists only of making bold conjectures and trying to falsify them. Second, evolution by natural selection is not a scientific theory, since it is not falsifiable. Popper (1978) eventually retracted the second claim, but in any case there are many logical and historical reasons why Popper's falsificationism should not be taken as the standard of scientific method (Sober, 2008; Thagard, 1988).

2. A follower of the historian of science Thomas Kuhn (1962) could say that Darwin's views are a new paradigm that is incommensurable with the old, creationist paradigm, so there is no objective way of choosing one over the other. In his later writings, Kuhn backed away from such a subjectivist interpretation of his earlier views, and Thagard (1992) argued that radical incommensurability and irrationality are not features of scientific revolutions.

3. Bayesians say that belief revision should be in accord with probability theory, in particular with Bayes's rule, which says that the probability of a hypothesis given evidence is equal to the result of multiplying the prior probability of the hypothesis times the probability of the evidence given the hypothesis, all divided by the probability of the evidence: $P(H/E) = P(H)*P(E/H) / P(E)$. As a theorem of probability theory, this rule is unassailable, but its application to actual cases of reasoning is highly problematic (Thagard, 2000). What is the correct interpretation of probability here, and how does one come up with numbers such as the probability of the evidence given Darwin's theory? Sober (2008) provides a sophisticated probabilistic discussion of the evidence for natural selection, but it would be hard to apply in science education.

4. Thagard (1988, 1992) argued that Darwin's theory has been and should be accepted because of its explanatory coherence, which takes into account how it compares with alternative theories such as creationism according to the range of facts explained, simplicity, and analogies such as Darwin's comparison of natural selection to artificial breeding. This account fits very well with Darwin's own argument and with widespread inferential practice (Kitcher, 1981).

If the explanatory coherence account of scientific justification is correct, it should be appropriate to try to convince students of Darwin's theory using inference to the best explanation: lay out the evidence and alternative explanations, and determine what competing hypotheses explain the evidence best. Unfortunately, attempts to use explanatory coherence for educational purposes revealed pedagogic problems (Schank & Ranney, 1992). Philosophers, scientists, and science educators may take it for granted what hypotheses and evidence are, but these concepts are by no means clear to average students. Hence we need an elemental exposition of scientific method starting with the clarification of the nature of evidence, explanations, hypotheses, and theories (see Thagard, 2010a,c). Only then can students have a chance of grasping that they should accept Darwin's theory, not just because it is in the textbook or class lectures, but because it is a powerful explanatory theory. Students need to learn not only what to believe, but how to believe for the right scientific reasons. Then getting to Darwin presupposes the methodological leap of getting to science first, a historically and psychologically difficult process. After all, the systematic evaluation of competing hypotheses with respect to experimental and observational evidence only began to be practiced and understood in the sixteenth and seventeenth centuries, and most students hit high school and even university with minimal exposure to it.

Coherence Difficulties

Suppose that students overcome conceptual difficulties and understand Darwin's theory, and suppose that they overcome methodological difficulties and grasp the importance of inferring the best explanation of all the evidence. It still does not follow that they will become avid Darwinians, if they see evolution as incoherent with alternative explanations that they grew up with. Christian, Muslim, and other religious traditions do not mention evolution by natural selection, and offer a strikingly different account of the origin of species: divine creation. In Darwin's time, creation was the dominant view even in scientific circles, and Darwin constructed *On the Origin of Species* explicitly as an alternative to creationist explanations. According to the Bible, God produced light, sky, water, land, vegetation, and animals over a few days, which is a very different story from the biological view that species on Earth evolved over the course of billions of years. Students who grow up with the religious story will need to either reject it or revise it dramatically before they become capable of accepting evolutionary explanations of the origin of species. Preston and Epley (2009) present evidence that the conflict between religious and scientific explanations can occur automatically, such that increasing the perceived value of one decreases the evaluation of the other. Later we will discuss various pedagogical strategies for dealing with the incoherence that students face between biological and theoretical explanations of how *Homo sapiens* and other species came to be.

The Darwinian, biological view of human origins is incoherent not only with religious views, but also with common psychological views that fit very well with religious accounts. According to many religious traditions, God gave people souls that enable them to make free decisions and to survive the death of their bodies. This doctrine contradicts the biological view that thought results from processes in brains that became increasingly complex through hundreds of millions of years of evolution. In order to accept the application of Darwin's theory to the human species, including its mental faculties, students may need to replace the religion-derived explanatory account of minds as souls with the relatively recent view of minds as brains. At the very least, students will need to revise their religious and psychological beliefs, as some scientists have, to allow for compatibility between religion and biology.

In sum, there are three major cognitive obstacles to accepting Darwin's theory of evolution by natural selection: understanding the theory in the face of conceptual difficulties, grasping the nature of scientific evaluation of evidence, and dealing with incoherence between biological explanations and much more familiar explanations of life and mind derived from religion and commonsense psychology. These alternative explanations are not only familiar, but also highly emotionally attractive, so that there are serious emotional obstacles to getting to Darwin.

Emotional Obstacles

Rationally, people are supposed to form their beliefs on the basis of evidence, not according to what they want to be true. But it would be naïve to ignore the fact that people often shape their beliefs in part by their goals, a process that psychologists call "motivated inference" (Kunda, 1990, 1999). For example, people may overestimate their chances of winning a lottery or underestimate their chances of getting a disease because of what they want to happen. Thagard (2006a, ch. 8) showed how such motivated inferences can develop from a process of emotional coherence, in which beliefs arise through a combination of explanatory coherence (as described above) and emotional influences: people are inclined to accept beliefs that fit with their goals, and to reject beliefs that conflict with their goals.

The problem that many people have with accepting Darwin's theory is not just that it conflicts with their existing religious and psychological beliefs, but that it conflicts with some of their deepest personal motivations. People do not just think that evolution by natural selection is false: They want it to be false. Even people who believe in evolution may admit that they would prefer it to be false (Brem, Ranney & Schindel, 2003).

Let us try to identify the key motivations that generate emotional obstacles to accepting evolution by natural selection. There are many appealing aspects of the religious-psychological perspective that God created the world and human souls. First, this perspective carries the comforting picture, found in Christianity and some other religions, of a loving and all-powerful God. Many people receive from this picture great reassurance that the many problems that life presents (disappointment, disease, death) are transitory parts of a beneficent, divine plan. Second, this divine plan potentially includes everlasting happiness, which requires the existence of an immortal soul that is immune from the threat of bodily death. Third, within our lifetimes, people are not completely constrained by the kinds of physical and biological forces that generated evolution, but rather operate by free will. Not only is free will supported by our subjective experience of having genuine choices to make, it also fits with our preferred view of ourselves and others as responsible agents. A major reason why people recoil from Darwinian approaches to biology and psychology is that they see it as rendering their lives meaningless, devoid of the spirituality and morality that make life worth living.

So students may see themselves not just as facing a cognitive choice between two competing theories of the origin of species, but also as facing an emotional choice between two competing systems of values. On the one hand, there is the familiar, reassuring religious picture that includes a caring God, immortality, free will, moral responsibility, and meaningful lives. On the other land, there is the gloomy scientific picture of humans as specks in the vast universe, irrevocably doomed to die after a brief life devoid of freedom, morality, and purpose. So it is small wonder that students are highly motivated to look skeptically at the vast evidence for Darwin's theory, and eager to seize on the feeble arguments of proponents of creationism and intelligent design. Who would want to accept Darwin if doing so would suck all meaning from life? (It does not have to: see the discussion below and Thagard, 2010a.)

Resistance to evolution by natural selection can also take on a political dimension. Acceptance of evolution varies wildly between different countries, partly as a result of political pressures. In the United States, for example, skepticism about Darwin is associated with highly conservative political views as well as religious fundamentalism. Politics also figure in other scientific controversies, such as whether the main cause of global warming is a human-made increase in carbon emissions. Resistance to claims that people are responsible for climate change tends to come from politicians sympathetic to the oil industry and antagonistic to government intervention (see chapter 5). Hence change in scientific beliefs can encroach on political values as well as on personal motivations.

In the next section, we will address how science educators might draw on philosophy and psychology to allay students' fears about the psychological devastation that appreciating Darwin might bring. Our purpose here is to emphasize that resistance to Darwinian ideas is not just a cognitive matter of comprehension, scientific inference, and belief revision, but also an emotional matter of reconciling conflicting views of what is valuable about life. When cognitive and emotional obstacles combine and interact, it is not surprising that evolution by natural selection faces enormous resistance, as it did in the nineteenth century and continues to do today. Conceptual change concerning topics that matter to people is often an emotional process as well as a cognitive one. So what is the science educator to do?

Implications for Science Education

Teachers who want their students to understand and accept Darwin's theory of evolution might pursue one of three pedagogic strategies: detachment, reconciliation, and confrontation. By detachment we mean that the presentation and discussion of evolution and other biological theories can be discussed with no mention of the coherence difficulties and emotional obstacles that arise from religious issues. Science is simply detached from religion, in line with the American doctrine of separation of church and state.

By reconciliation we mean that possible conflicts between science and religion are mentioned, but they are resolved by finding compatibilities between scientific and religious views. Several avenues of reconciliation are possible, including a postmodernist assertion that science and religion are just different modes of discourse, a separatist assertion that science and religion are not incompatible because they operate in different realms, and a quasi-scientific advocacy of a view called "theistic evolution," according to which God is the designer of the evolutionary process.

Finally, at the most extreme, there is the strategy of confrontation, in which scientific views are presented as contradictory with and superior to the religious views that provide the cognitive and emotional motivations for rejection of evolution by natural selection. Which of these strategies is the best for different kinds of science education depends on a host of philosophical, scientific, psychological, and political factors.

Politically, by far the easiest pedagogic strategy is detachment. High school and university teachers can simply present the theory of evolution by natural selection, elucidating its conceptual complexity and demonstrating its enormous explanatory power. Detaching this presentation from religious issues avoids any kind of conflict with religious issues, eliminating the expenditure of emotional energy needed to deal with broader philosophical issues that science educators may be poorly equipped to address.

Detachment from religious issues may well be the best strategy for most science teachers, but it comes at a substantial intellectual cost. Perhaps the science educator could be satisfied with having students understand Darwin's theory, regardless of whether they actually believe it. But students are unlikely to expend the substantial effort required to overcome the conceptual difficulties that impede understanding if they are convinced that it is false (Sinatra et al., 2003). If such convictions arise because of coherence difficulties and emotional obstacles arising from students' common perceptions that Darwin conflicts with their theological beliefs and personal motivations, then the detachment strategy will have limited pedagogical effectiveness. The educational goal of getting students to appreciate the scientific value of the theory of evolution by natural selection would not be fully accomplishable using the detachment strategy. Moreover, detachment is no help in dealing with political attempts in some jurisdictions to require that creationism be taught in schools as an alternative to Darwin's theory.

So it might be better instead to address the perceived conflict between science and religion explicitly and attempt to remove it by one of various approaches. The postmodernist approach to reconciliation is highly relativist, asserting that there is no real competition between Darwinian and creationist accounts because they are just different ways of talking about the world, different language games, different modes of discourse. If science and religion are both just social constructions, they need not be understood as making different substantive claims about the origins of species, so there is no reason to see a conflict between them.

This view of science is held by many sociologists and historians of science, and has even been advocated by some science educators. However, there are good philosophical reasons for rejecting it (see, e.g., Thagard, 1999; Brown, 2001); and the postmodernist views of science and religion are unlikely to be accepted generally either by practicing scientists or by religious leaders. Science purports to make claims about the universe, for example, that it is about 13.7 billion years old, that life has been evolving on the planet Earth for more than 4 billion years, and that the human species is a result of this evolutionary process. There is substantial empirical evidence and theoretical basis for each of these claims, which markedly differentiates them from fictional modes of discourse such as Harry Potter novels. Hence the postmodernist, relativist approach to reconciling science and religion is unlikely to be intellectually effective.

Another way of attempting to reconcile science and religion is to assign them to separate realms of concern, what Stephen Jay Gould (1999) called "magisteria." On this view, science rules in the realm of fact and evidence, but religion has much to contribute to issues of morality and meaning. This view may be reassuring to some, but it fails to deal with conflicts over specific claims such as how the human species came to be. Moreover, separatism with respect to realms does not address the crucial question of how religion could tell us much about morality and meaning if its fundamental assumptions, such as that the universe was created by a caring God, are false.

The currently most popular way of reconciling Darwinism and theology is sometimes called "theistic evolution." This approach pushes the role of God back before the design of species to the creation of the universe (see, e.g., http://www.theisticevolution.org/). On this view, God did not create individual species in the way that the Bible describes, but rather designed the universe in such a way that physical changes and biological evolution would eventually lead to the production of the human species. This reconciliation is currently favored by the Roman Catholic Church, mainstream Protestant denominations, and most forms of Judaism. Politically and psychologically, theistic evolution may be a good strategy for reassuring threatened students that Darwin is not incompatible with flexible forms of theology that do not require that the Bible be literally true.

There are two main drawbacks, however, to using the presentation of theistic evolution as a strategy for science education. First, it will not work for the fundamentalist students most in need of help with coherence difficulties and emotional obstacles to accepting Darwin. Such students will find theistic evolution almost as repugnant as Darwin's theory, because it violates their deep faith in the literal truth of their favorite religious texts and the ongoing intervention of God in human lives. Second, although theistic evolution avoids direct conflict with biological evolution, it still methodologically violates the general scientific approach of evaluating explanatory theories with respect to all the available evidence. There is no evidence that God created the universe originally, let alone that it was designed in such a way that after more than 13 billion years an intelligent species would emerge on the third planet of a minor star. Theistic evolution may not directly contradict biological evolution, but it clangs dramatically with the general scientific approach to understanding the origins of the universe, which can be addressed using the resources of physics (Steinhardt & Turok, 2007).

If relativism and reconciliation fail, there remains the strategy of a more general confrontation between science and religion, subsuming the specific dispute between evolutionists and creationists. Polemicists such as Dawkins (2006) and Dennett (2006) have pursued such confrontation, arguing that there are scientific grounds for rejecting all the fundamental claims of religion such as the existence of God. If science supersedes religion in general, then there is no particular conflict between Darwin's theory of evolution and the cognitive claims and emotional offerings of theology. Religion is false, so it ought to pose no threat to scientific theories supported by evidence, including natural selection. Pursuing this line of argument, however, may be difficult in places where criticism of religion is politically or socially prohibited.

To be effective, the confrontation between science and religion needs to address cognitive and emotional issues that tend to be neglected by advocates of science. Cognitively, they tend to assume that the evidence-based methods of science supersede the faith-based thinking of religious people, but religionists reply that the preference for scientific evidence is just another kind of faith. Thagard (2010a, ch. 1) argues that there are good reasons for preferring evidence over faith as a way of fixing beliefs. One major reason is that faith provides no way of adjudicating between competing claims such as whether the true divinity is the god of the Christian Bible or the myriad gods of the Hindu, Roman, and Greek traditions. In contrast, science proposes the procedure of evaluating competing explanatory hypotheses with respect to all the evidence, choosing the best. This procedure offers a way of settling intellectual disputes that is far more effective and far less rancorous than aggressive assertions about whose faith is superior. Moreover, scientific methods of figuring out how the world works have had amazing practical effects, seen in technologies such as lasers and computers. Hence science can legitimately be taken to supersede religion when cognitive conflicts about facts and theories occur.

What about emotional conflicts concerning values? When we discussed the emotional obstacles to accepting Darwin, we described the appealing package that comes with religion, including a caring God, immortality, free will, meaning, and morality. Thagard (2010a) argues at length that evidence or lack thereof requires abandoning the first three of these appealing ideas. Nevertheless, meaning and morality can survive in naturalistic forms consistent with human biology and psychology, without God, immortality, or free will. Thagard presents evidence for concluding that the meaning of life is love, work, and play, which together meet fundamental human psychological needs. Contrary to the widespread belief that morality requires religion, a moral system can also be developed by taking into account scientific evidence about psychological and biological needs. If these claims are correct, then at least some of the emotional appeal that leads to the rejection of Darwinian and scientific ideas can be countered by developing scientifically informed philosophical views of meaning and morality, in the long tradition of naturalists such as Epicurus and John Stuart Mill.

We concede, however, that the confrontational strategy is not well suited to many educational situations. It is much more likely to be pedagogically effective in advanced university courses than in high school classes, especially ones occurring in bastions of religious fundamentalism such as Arkansas and Alberta. Confrontation may be politically difficult if it leads to major conflicts between teachers, students, and parents. It may not even be the psychologically most effective way of bringing about emotional conceptual change, for which more subtle methods of persuasion may be less likely to lead to disputants digging in and dogmatically asserting their positions. Much more research is needed into how belief revision and emotional change interact with each other.

We leave it to science educators to figure out which strategies for dealing with the cognitive and emotional conflicts between Darwinism and religion are most likely to be effective in their particular circumstances. There are solid intellectual grounds for rejecting the relativist and separatist approaches to reconciliation, but the theistic-evolution approach may be a useful transition in assisting students to see how science might proceed without generating a direct conflict between evolution and theology. Often, however, science educators will have to fall back on the local strategy of detaching science from religion, even if disputes on a broader level may require dealing with a confrontation between conflicting cognitive and emotional stances. Ironically, political attempts to keep creationism out of the schools may undermine cognitive strategies that help students appreciate the strengths of Darwinism by contrasting it with theological approaches.

Any effective strategy for science education should include a more sophisticated understanding of scientific methodology than is usually present in science courses. Science textbooks often begin with a simplistic view of scientific method as hypothesis testing: generate a hypothesis, design an experiment to test it, and reject the hypothesis if the experimental prediction is not observed. The methodology of Darwin and other scientists is much more complex, including processes such as dealing with experimental anomalies and evaluating theories by comparing them with competing theories. Thagard (2010c) offers a concise introduction to issues in the philosophy of science relevant to controversies about evolution.

Conclusion

Darwin's theory of evolution by natural selection is one of the most important intellectual developments in human history. An understanding of it should be part of the mental equipment of science students and indeed of all educated people. Students ought to be taught to appreciate its power in explaining the origin and distribution of species, as well as in explaining many aspects of human existence such as the biological basis of our perceptual, emotional, and cognitive capabilities. Ideally, students should also acquire an appreciation of the limitations of Darwinian explanations, evident in overextensions to cultural phenomena proposed by some sociobiologists and evolutionary psychologists (Richardson, 2007). Some of the feared consequences of Darwin's ideas, for example, that people are genetically determined by their selfish genes toward greed, violence, and destruction, are based on misunderstandings of evolutionary theory. Natural selection can explain how desirable human characteristics such as altruism and empathy could have evolved through biological processes such as group selection and neural development (Sober & Wilson, 1998; Thagard, 2010a).

We have reviewed what we see as the main cognitive and emotional obstacles to the more widespread acceptance of Darwinism. We hope that identifying them may help to develop more effective teaching strategies for overcoming resistance to the powerful scientific theory of evolution by natural selection.

15   Acupuncture, Incommensurability, and Conceptual Change

Paul Thagard and Jing Zhu

Introduction

In November 1997, the U.S. National Institutes of Health (NIH) conducted a consensus development conference on acupuncture. Like the previous 106 consensus conferences sponsored by NIH since 1977, the acupuncture conference consisted of presentations to a panel charged with making recommendations concerning medical practice. But it was unusual in being the first cosponsored by the new NIH Office of Alternative Medicine, and the first to consider therapies from outside the Western medical tradition. Acupuncture, involving the insertion of needles under the skin at prescribed positions, is a central component of the 2,500-year old system of traditional Chinese medicine. This system of medicine employs a conceptual system very different from the one that has evolved in Europe and America over the past century and a half. Allchin (1996, p. S107) says that the contrasting traditional Chinese and Western views of acupuncture "offer a particularly deep version of Kuhnian incommensurability."

In the 1960s, philosophy of science was scandalized by the suggestion that competing theories might be incommensurable with each other (Kuhn, 1962; Feyerabend, 1965). If two conceptual systems such as those comprising the oxygen and phlogiston theories of combustion are radically different, then rational comparison and assessment of them becomes difficult if not impossible. Subsequent discussions have shown that claims of radical incommensurability in the history of science were greatly exaggerated (see, e.g., Laudan, 1984; Nersessian, 1984; Thagard, 1992). Although competing scientific theories may indeed occupy very different conceptual systems, there is usually enough conceptual, linguistic, and evidential overlap that rational assessment of the comparative explanatory power of the two theories can take place. In Kuhn's most recent writings, incommensurability is no longer a dramatic impediment to the comparability of theories, but rather an unthreatening observation on difficulties of translation and communication (Kuhn, 1993).

Nevertheless, even if rationality-destroying incommensurability is rare within science, it would not be surprising if it occurred at the boundaries between science and nonscience. Contrast, for example, the cosmology of modern astrophysics with that of Australian aboriginals. The ontology, concepts, and linguistic context of these cosmologies are so radically different that explicit comparison and rational evaluation in terms of the standards of each will be very difficult. Of course, from the empirical and theoretical perspective of Western science, the superiority of astrophysics over aboriginal cosmology is obvious, as are the advantages of the scientific perspective. Still, it is possible that radical incommensurability, of the sort that Kuhn and Feyerabend mistakenly attributed within Western science, may arise between science and alternative views of the world.

This chapter is an investigation of the degree of incommensurability between Western scientific medicine and traditional Chinese medicine, focusing on the practice and theory of acupuncture. In the next section, we very briefly sketch the conceptual and explanatory structure of modern medicine. We then provide a more detailed description of the structure of traditional Chinese medicine, oriented around such concepts as *yin*, *yang*, *qi*, and *xing*, which enables us to show how the conceptual and explanatory differences between Western medicine and traditional Chinese medicine generate impediments to their comparison and evaluation. We outline linguistic, conceptual, referential, and explanatory difficulties that might be taken to imply that traditional ideas about acupuncture are incommensurable with Western medicine. We argue, however, that the difficulties can to a large extent be overcome, as they were at the NIH meeting, in service of an attempt to improve medical treatments. Our conclusion is that the dramatic differences between Western and traditional Chinese medicine do not provide insurmountable barriers to rational evaluation of acupuncture.

One positive contribution of this chapter is that its display of the conceptual and explanatory structure of traditional Chinese medicine provides an informative contrast that highlights aspects of the nature of Western science. We conclude with a discussion of how an appreciation of different medical frameworks can require conceptual change that is both intentional and emotional. Thinkers are more likely to appreciate an alternative conceptual scheme if they have the goal of gaining such an appreciation, and acquiring the alternative scheme may involve changes in emotional attitudes as well as in concepts and beliefs.

Western Scientific Medicine

What is the structure of modern medicine? Biomedical theories are not naturally represented as formal axiom systems, but can naturally be characterized in terms of hierarchical cognitive structures (Schaffner, 1993; Thagard, 1999). Figure 15.1 depicts at a very general level the conceptual and explanatory structure of scientific medicine as it has evolved since Pasteur proposed the germ theory of disease in the 1860s. Diseases can be classified according to the bodily systems that they affect, for example, as heart or skin diseases, but a deeper classification is based on the causes of disease. Modern medicine recognizes four kinds of causes of disease: infectious agents such as viruses, nutritional deficiencies such as lack of vitamin C, molecular-genetic disorders such as cancer, and autoimmune reactions such as the attack on the connective tissue that produces lupus erythematosus.

For each class of disease, there is an explanation schema that specifies a typical pattern of causal interaction. Explanation schemas and similar

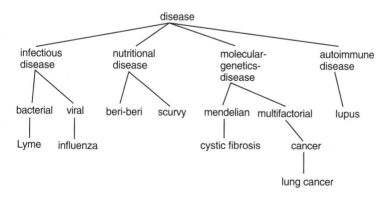

Figure 15.1
Hierarchical organization of disease explanations, with examples of particular diseases. From Thagard (1999).

abstractions have been discussed in philosophy and cognitive science using varying terminology; see, for example, Darden and Cain (1989), Giere (1999), Kelley (1973), Kitcher (1981, 1993), Leake (1992), Schaffner (1993), Schank (1986), and Thagard (1988, 1992).

Infectious diseases fall under the following explanation schema, which became very successful in the nineteenth century:

Germ Theory Explanation Schema

Explanation target:

Why does a **patient** have a **disease** with **symptoms** such as fever?

Explanatory pattern:

The **patient** has been infected by a **microbe**.

The **microbe** produces the **disease** and **symptoms**.

To apply this schema to a particular disease, we need to replace the terms in boldface with specific examples or classes of examples. For example, influenza instantiates the schema by specification of symptoms such as fever, aches, and cough and by specification of the class of flu viruses that cause the symptoms. Figure 15.2 diagrams the causal structure of the germ theory of disease. Germs such as viruses and bacteria cause infections that produce symptoms that develop over time, constituting the course of the disease. Treatments such as antibiotics and vaccines can kill the germs or inhibit their growth, thereby stopping or preventing the infection that produces the symptoms.

Analogous explanation schemas for nutritional, molecular-genetic, and autoimmune diseases have been presented elsewhere (Thagard, 1999).

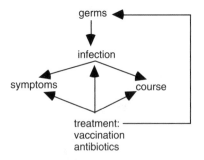

Figure 15.2
Causal structure of the germ theory of disease. From Thagard (1999).

Here, we present only enough of the conceptual and explanatory structure of modern medicine to provide a contrast with an alternative approach.

Traditional Chinese Medicine

Traditional Chinese medicine developed for more than 2,500 years almost entirely free from Western influences. Its theories and practices of diagnosis and treatment are remarkably different from those of Western medicine. Some of its treatments such as herbal therapies and acupuncture are currently receiving increasing attention in the West, part of a rapidly growing interest in alternative and complementary medicine. As we will describe later, numerous experimental and clinical studies have confirmed that acupuncture treatments can relieve pain and reduce nausea. However, the theories of traditional Chinese medicine seem bizarre from the point of view of modern Western medicine. Even if some proponents of traditional Chinese medicine claim that Western and Chinese medicine should complement each other, they admit that Chinese medicine is organized on totally different principles (Porkert & Ullmann, 1988, p. 55). In this section, we will try to display this organization and outline the conceptual and explanatory structure of traditional Chinese medicine.

The Balance of Yin and Yang

In ancient China, people believed that everything in the universe consists of two opposite but complementary aspects or forces, which combine to create a whole unit (Zhen, 1997). *Yin* and *yang* refer to the two basic categories of the universe, negative and positive respectively, and they are in constant flux. Every thing or event in the world is to be regarded as the interaction of an active and a conservative force, each of which has its own peculiar characteristics that determine the nature of the thing or event. According to *Yellow Emperor's Classic of Internal Medicine: Plain Questions*, one of the most important and original classics of traditional Chinese medicine, "the principle of *yin* and *yang* is the way by which heaven and earth run, the rule that everything subscribes, the parents of change, the source and start of life and death" (Guo, 1992, ch. 5).

The original meaning of the two words *yin* and *yang* in Chinese referred, respectively, to the side of a mountain that lies in shadow and the side that lies in sun. *Yin* could also refer to the shaded bank of a river, *yang* to the sunlit bank. But the terms are no longer strictly confined to their original meaning and have become basic and abstract categories in both Chinese philosophy and people's ordinary thinking. Typically, dynamic, positive, bright, warm, solid features are defined as *yang*, while static, negative, dark, cold, liquid, and inhibiting features are characterized as *yin*. Sunlight and fire are hot while moonlight and water are cool, so the Sun and fire are *yang* while the Moon and water are *yin*. *Yin* and *yang* are complementary to and interdependent on each other, even though they are opposites. For every individual thing, the *yin* and *yang* it contains do not remain in a static state, but are constantly in a dynamic equilibrium affected by the changing environment.

Like everything else, the human body and its functions are all governed by the principle of *yin* and *yang*. Remaining healthy and functioning properly require keeping the balance between the *yin* and *yang* in the body. Diseases arise when there is inequilibrium of *yin* and *yang* inside the body. This principle is central to traditional Chinese medicine, and its application dominates the diagnosis, treatment and explanation of diseases. For example, a patient's high fever, restlessness, a flushed face, dry lips, and a rapid pulse are *yang* symptoms. The diagnosis will be a *yin* deficiency, or imbalance brought by an excess of *yang* over *yin*. Once the *yin-yang* character of a disease is assessed, treatment can restore the balance of *yin* and *yang*, for example, by using *yin*-natured herbs to dampen and dissipate the internal heat and other *yang* symptoms. The imbalance of *yin* and *yang* can be caused by either exogenous factors, such as climate, traumatic injuries, and parasites, or endogenous factors, such as extreme emotional changes (anger, melancholy, anxiety, and so on), abnormal diet, intemperance in sexual activities, and fatigue. Figure 15.3 displays the structure of the causal network underlying the *yin-yang* explanation of disease. As with the germ theory of disease shown in figure 15.2, causes produce a set of symptoms and their course of development.

The way in which this causal structure explains disease can also be described by the following schema:

Yin and *Yang* Balance Theory Explanation Schema

Explanation target:

Why does a **patient** have a **disease** with associated **symptoms**?

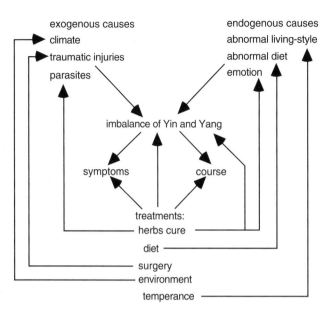

Figure 15.3
Causal structure of disease concepts in the theory of *yin* and *yang* balance.

Explanatory pattern:
The **patient's** body is subject to exogenous and endogenous **factors.**
The **factors** produce an **imbalance** of *yin* and *yang.*
The **imbalance** of *yin* and *yang* produces the **disease** and **symptoms.**

This is the most general and fundamental pattern of disease explanation in traditional Chinese medicine, but there are also some more specific explanation schemas.

The Theory of the Five Xing

According to Aristotle and Hippocrates, everything in the world consists of four fundamental elements: earth, air, fire, and water. Similarly, the ancient Chinese considered metal, wood, water, fire, and earth to be fundamental. In Chinese, each of these is a *xing,* and the five collectively are called five-*xing.* There are two important differences between the theory of five-*xing* and the theory of four elements in ancient Greece. First, even though the five *xings* were considered as the basic components of the universe, the ancient Chinese did not use them to analyze the substantial constitution of particular things. Rather, the five *xings* are five basic categories that can be used to classify things according to their properties and relationships to other things. Second, the five *xings* are not independent of each other, but have significant relationships and laws of transformation among them. Hence, instead of translating *xing* as "element," various commentators prefer to call five *xings* the "Five Transformation Phases" or the "Five Phases of Change" (Porkert & Ullmann, 1988; Unschuld, 1985).

There are two basic kinds of relation or sequence among five *xings*: Mutual Promotion (Production) and Mutual Subjugation (Conquest). The principle of Mutual Promotion says that five *xings* may activate, generate, and support each other. It is through these promotions of the elements that five *xings* continue to survive, regenerate, and transform. The sequence of Mutual Promotion is as follows: wood promotes fire, fire promotes earth, earth promotes metal, metal promotes water, water promotes wood, and wood again promotes fire. The principle of Mutual Subjugation concerns relations such as restraining, controlling, and overcoming. Mutual restraint keeps the balance and harmony among the five *xings*. Wood subdues earth; earth subdues water; water subdues fire; fire subdues metal; metal subdues wood; wood in its turn acts on earth. Figure 15.4 shows the Mutual Promotion and Mutual Subjugation relationship among the five *xings*.

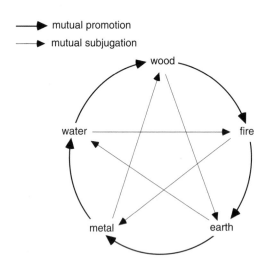

Figure 15.4
Mutual Promotion and Mutual Subjugation relations among five *xings*. Adapted from Shen and Chen (1994), p. 17.

The meaning of the principles derives from experience. Fire is created when wood is burned. Ash (earth) is left after burning. All metals come from earth and liquefy on heating, while water is indispensable for growing trees and vegetation. These relations support the principle of Mutual Promotion (Production). On the other hand, the ancient Chinese noticed that trees grow in the earth, impoverishing the soil. To prevent floods, dams and channels are built with earth. Water puts out fire, while metals can be softened and melted by fire. A sword or ax made of metal can be used to fell a tree. These relations are summarized by the principle of Mutual Subjugation (Conquest).

Most things in the world can be classified into one of the five basic categories according to their properties, functions, and relations with others. For example, the liver is similar to wood with respect to its mild features, and the heart warms the whole body so it is analogous to fire. The spleen is responsible for assimilation of nutrients and corresponds to the earth. The lung is clear, analogous to metal. The kidney is similar to water by virtue of its responsibility of regulating fluids in the body.

In diagnosis and treatment, those things classified into the same kind are related to each other and have the same mutual relations with the objects in the neighboring categories. For example, a disease in the liver calls attention to the eyes, tendons, and the emotion of anger. Great anger is considered very harmful to the liver in traditional Chinese medicine. The liver pertains to wood, which flourishes in spring, so that liver diseases are prevalent in spring. The classification and correspondence in terms of five *xings* illustrate the mutual relationship between the human body, the seasons, climate factors, senses, and emotions. According to the principles of Mutual Promotion and Subjugation, a disease in one organ is not isolated from the other organs. A disease in the liver (wood) is probably due to a functional deficiency of the kidney (water), so the treatment should not only be aimed at the liver, but should also enhance the function of kidney as well as those of others. Thus the perspective of traditional Chinese medicine is more holistic than the Western perspective, which tends to look for the seat of a disease in a particular organ.

The theory of five *xings* specifies aspects of the more general theory of *yinyang* balance. The improper function of an organ is originally caused by an imbalance of *yin* and *yang*, and in turn influences the harmony between other organs, which can also be analyzed in relation to *yin* and *yang*. Here is the explanatory schema:

Five *Xings* Explanation Schema

Explanation target:

Why does a **patient** get a **disease** with associated **symptoms?**

Explanation pattern:

The **imbalance** of *yin* and *yang* causes one or more organs, which belong to the corresponding *xing*, to **malfunction.**

The **malfunction** of one organ produces the **disorder** among all the organs, which are related between each other according to the rules of the theory of five *xings*.

The **disorder** among organs produces the **disease** and **symptoms.**

The Circulation of *Qi*

Another fundamental concept in traditional Chinese medicine is *qi*, which plays a central role in the theoretical background to such therapies as acupuncture, moxibustion, and massage. In ordinary Chinese, the term *qi* refers mostly to air or gas, and sometimes is also used to indicate a kind of emotion—anger. In the terminology of Chinese medicine, *qi* has a different meaning. First, *qi* is not a type of substance and has no fixed shape or constitution. Second, it is indispensable for life. Third, it is responsible for the resources of the function and operation of organs and the whole body. *Qi* has variously been interpreted in terms of the Greek *pneuma*, vital force, or energy (Lloyd, 1996, Lu & Needham, 1980). *Qi* cannot be observed directly, but with long and assiduous training and practice, a doctor can supposedly detect its flow and changes in a patient. A person can also detect the flow of *qi* and control its direction in some degree by exercises and meditation (Moyers, 1993).

There are basically two kinds of *qi*, *congenital qi* inherited from one's parents and vital for one's life, and the other type acquired after birth. We get the acquired *qi* from food and water, which is assimilated by the spleen and stomach, and from the air inhaled by the lungs. The acquired *qi* is constantly replenished, and is fundamental to maintaining the life activities of the body. Because *qi* is dynamic, active, and warms the body, it falls under the *yang* category. Blood and body fluids, two kinds of fluids circulating inside the body, have the functions of nourishing and moistening. Therefore they belong to the *yin* category. *Qi* is capable of producing and controlling blood, warming and nourishing the tissues, and activating the functions of organs.

Qi circulates along channels within the body called "meridians." The system of meridians is unique to traditional Chinese medicine regarding the human body, and does not correspond to blood vessels or nerves. The resource of *qi* inside meridians comes from the internal organs, such as the heart, spleen, lung, and stomach. As a unit, the system works to reinforce the coordination and balance of bodily functions.

Disease occurs when the circulation of *qi* is obstructed. Doctors need to identify where and why the flow of *qi* is blocked and carry out the proper treatment to restore the circulation of *qi*. Deficiency of *qi* can also cause illness, and the appropriate treatment is to replenish it. Figure 15.5 displays the causal structure of the theory of *qi*, and we can summarize disease explanations in the following schema:

Theory of Qi Explanation Schema

Explanation target:
Why does a **patient** have a **disease** with associated **symptoms**?

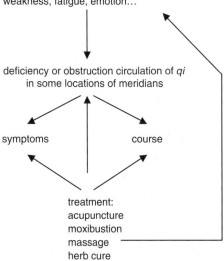

disorder or malfunction of organs
disharmony of the interaction between *qi* and blood
weakness, fatigue, emotion…

deficiency or obstruction circulation of *qi*
in some locations of meridians

symptoms   course

treatment:
acupuncture
moxibustion
massage
herb cure

Figure 15.5
Causal structure of diseases due to *qi* blockage.

Explanation pattern:

The body of the **patient** contains a meridian system that conducts the flow of *qi*.

An **obstruction** occurs that blocks the flow of *qi*.

The *qi* blockage produces the **disease** and **symptoms**.

Before proceeding to discuss philosophical issues concerning the relation of Western and Chinese medicine, we should stress that both these traditions are concerned primarily with the treatment of patients, and that the development of explanatory theories of diseases has been driven largely by this practical aim.

Incommensurability

It is obvious from our brief review that the conceptual and explanatory structure of traditional Chinese medicine is very different from that of Western medicine; but are these differences so large that the two systems cannot rationally be compared? This question is of considerable current practical importance because of live controversies concerning the medical legitimacy of acupuncture and other traditional Chinese treatments. If traditional Chinese medicine and Western medicine are mutually unintelligible, then evaluation of one within the framework of the other would seem to be impossible. We will now consider four potential impediments to mutual intelligibility: the linguistic differences between Chinese and Western languages, the differences in conceptual organization between Chinese and Western systems of thought, referential differences between Chinese and Western theories, and explanatory differences involving notions of causality and correspondence.

Our discussion will distinguish between "strong" and "weak" incommensurability. Two theories or conceptual schemes are strongly incommensurable if they are mutually unintelligible, so that someone operating within one conceptual scheme is incapable of comprehending the other. Weak incommensurability, however, does not imply mutual unintelligibility, but only that the two conceptual schemes cannot be translated into each other. If traditional Chinese medicine were strongly incommensurable with Western medicine, there would be no possibility of rational evaluation of Chinese medicine from the Western perspective. We shall argue, however, that the weak incommensurability that holds between the two medical traditions does not prevent rational evaluation of practices such as acupuncture.

Linguistic Differences

In both its spoken and written structure, the Chinese language is very different from European languages such as English, French, and German. Crucial terms from traditional Chinese medicine are not merely technical terminology, but are embedded in much broader linguistic usage. There are no terms in European languages that correspond even roughly to *yin* and *yang*, which are accordingly left untranslated. The term *qi* is often translated as "energy," but this translation is misleading if it generates an association with Western scientific concepts of electrical or mechanical energy, rather than with concepts such as breath, emotion, and force. Similarly, the translation of *xing* as "element" both adds and subtracts from the meaning of the Chinese term, since it adds the association of element as a fundamental constituent of the world, and loses the relational aspects of five-*xing* that are crucial to their explanatory roles.

The difficulty of translating Chinese medical terminology into European languages does not, however, show that Chinese and Western medicine are incommensurable. Even though there is no simple mapping of terms like *ying*, *yang*, *xing*, and *qi* into English, the fact that the linguistic divergence can be systematically described by writers such as Lloyd (1996), Porkert and Ullmann (1988), and Unschuld (1985) shows that comparison can proceed despite complexities of translation.

It is possible, however, that difficulties of translation run deeper than a lack of corresponding terms for *yin*, *yang*, and so on. Bloom (1981) argues that the structure of the Chinese language is radically different from European languages in that it lacks distinct markings for counterfactual conditionals, which are if-then statements in which the if-proposition is false. So it is not possible to make in Chinese such utterances as "If the Chinese government were to pass a law requiring that all citizens make reports of their activities to the police hourly, then what would happen?" He claims that Chinese speakers tend to brand the counterfactual as in some sense "unChinese." Bloom's claim is an instantiation of the linguistic relativity hypothesis of Whorf (1956), according to which differences in language generate radically different patterns of thought. If Bloom is right that the Chinese language enforces a non-Western attitude toward counterfactual conditionals, there may be differences in understanding of causality, since causation involves counterfactual dependence between events. It is possible, therefore, that the explanatory claims of Chinese medicine are untranslatable into European languages because they presuppose a very different conception of causality.

Bloom's linguistic claims have, however, been strongly challenged. Cheng (1985) argued that the psychological experiments that Bloom used to support his claim of linguistic divergence were methodologically flawed and used poorly translated materials; and his experiments did not replicate (Au, 1983). The Chinese language does in fact allow the statement of counterfactual conditionals, so there is no evidence of linguistic differences in the understanding of causality between Chinese and Western culture. Native speakers of Chinese, including the second author of this paper, report that Chinese people can in fact understand counterfactuals very well. In sum, the substantial linguistic differences between the Chinese and European languages do not generate insurmountable barriers to comprehension and translation, and therefore do not support claims of incommensurability between Chinese and Western medicine.

Conceptual Differences

The problem of comparing Western and Chinese medicine is not just that the terms are different, but that the concepts are differently placed in the conceptual hierarchical organization. Much current work in cognitive science views concepts as being organized in terms of kind hierarchies and part hierarchies (Thagard, 1992). For example, a chicken is a kind of bird, which is a kind of animal, and a beak is a part of a bird. There are problems, however, in placing Chinese concepts into the kind hierarchy of the Western system. *Yin* and *yang* do not seem to be kinds of anything familiar to Western thought. They are not things or substances or events or processes, and they involve a kind of abstraction not found in Western concepts. So the problem of translating these Chinese terms into English is not just a matter of finding the right term, but also reflects the fact that they do not fit into the Western conceptual organization.

Similar problems arise with *xing* and *qi*. The standard translation "elements" suggests that the five-*xings* are a kind of thing or substance, but this classification fits poorly with their crucial relations of promotion and subjugation. *Qi* would seem to be a kind of process, like energy, but its association with breath and force suggest that it is not a kind of process familiar to Western science or common sense. Thus the terms *yin, yang, xing,* and *qi* all represent concepts that do not fit naturally in the Western hierarchy of kinds. Moreover, the kind hierarchy for diseases in Chinese medicine tends to divide them into ones caused by too much *yin,* too much *yang,* or by blockage of *qi,* rather than according to the Western classification in terms of infectious, nutritional, molecular-genetic, and autoimmune causes. Thus the differences between Western and Chinese medicine are conceptual (mental) as well as linguistic (verbal).

Additional differences are found in the part hierarchies of the two systems. Traditional Chinese medicine did not permit autopsies and dissections, and so did not develop the detailed system of anatomy and physiology that evolved in Europe after the sixteenth century. According to Shen and Chen (1994), the term *zang-fu* refers to the five solid *zang* organs of the human body (heart, liver, spleen, lungs, and kidneys) and the hollow viscera (gall bladder, stomach, small intestine, large intestine, and bladder, and the *sanjiao*). The latter does not correspond to any part recognized in Western medicine, but consists of cavities of the chest and abdomen that are thought to be important for the flow of *qi.* The function of other organs is sometimes described in ways similar to Western medicine, but is sometimes radically different; for example, the heart houses the mental faculties. Traditional Chinese medicine ignores some organs such as the pancreas that are viewed as medically important in Western medicine.

Although the kind hierarchies and part hierarchies of traditional Chinese medicine and Western medicine are obviously different, it would be an exaggeration to say that they are mutually unintelligible. Concepts such as *qi* and *sanjiao* are undoubtedly alien to Western medicine, but their meaning can be acquired contextually from works such as Shen and Chen (1994). Conversely, practitioners of traditional Chinese medicine can acquire Western concepts such as *germ, virus,* and *pancreas.*

Buchwald and Smith (1997, p. 374) present a precise characterization of incommensurability that they report as Kuhn's final thoughts on the subject:

If two scientific schemes are *commensurable,* then their lexical structures can be fit together on one of the following two ways: (1) every kind, taxonomic or artifactual, in the one can be directly translated into a kind in the other, which means that the whole of one structure is isomorphic to some portion of the other; or (2) one structure can be grafted directly onto the other without otherwise disturbing the latter's existing relations. In the first case one scheme is subsumed by the other. In the second, a new scheme is formed out of the previous two, but it preserves intact all of the earlier relations among kinds. If neither case holds, the two systems are *incommensurable*.

Our discussion so far makes it clear that traditional Chinese medicine does not fit with Western medicine in the first way, since there is no direct translation of *yin, yang*, and kinds of Chinese disease into Western terminology. The second way of fitting does not work either, because grafting the two schemes together would require diseases to be classified simultaneously in conflicting ways, for example, as both infectious and caused by excessive *yin*. Thus on the characterization of Buchwald and Smith, Chinese and Western medicine are incommensurable, although the lack of fit does not imply that they are not comparable or mutually intelligible. This is weak incommensurability arising from untranslatability, in contrast with the strong incommensurability discussed by Laudan (1990, p. 121), who says that two bodies of discourse are incommensurable if the assertions made in one body are unintelligible to those utilizing the other.

Referential Differences

The meaning of a concept is a matter of both its relation to other concepts and its relation to the world. So far, we have been discussing linguistic and conceptual differences between Western and traditional Chinese medicine, but it is also clear that the two approaches make very different claims about the world. Not only are *yin, yang, qi*, and five-*xing* not part of the ontology of Western science, they are not even kinds of entities, properties, or processes that are part of that ontology. Conversely, traditional Chinese medicine does not even consider many of the referential claims of Western science, for example, concerning such entities as disease-causing microbes. A Kuhnian would be tempted to say that Western physicians and traditional Chinese doctors live in different worlds.

There is, however, considerable overlap in the two ontologies. Both Western and traditional Chinese physicians examine peoples' bodies with similar perceptual systems, even if there are differences in some examination techniques. Pulse taking is different in the two cultures, in that traditional Chinese doctors aim to detect pulses with three different grades of force, but both Chinese and Western doctors grasp wrists and detect pulses. Despite their different beliefs about what exists, it would be an exaggeration to place the traditional Chinese and Western doctors in different worlds.

Explanatory Differences

In Western scientific medicine, explanations are based on causal relations. A disease explains symptoms because the disease causes the symptoms, and the treatment is only judged to be effective if the treatment causes the elimination of disease. Although much of Western medicine is still based on the clinical experience of physicians rather than on scientific experiments, there is increased pressure to evaluate treatments using randomized, blinded, controlled trials (Sackett et al., 1996). Carefully controlled experiments are needed to determine whether treatments are causally effective, because they help to rule out alternative causes such as expectations and biases in physicians and patients (Thagard, 1999, 2010a).

According to Lloyd (1996, p. 113), traditional Chinese medicine similarly is interested in identifying causal factors, but it also has an additional explanatory style based on "correspondences." Unschuld (1985, p. 52) describes the role in ancient Chinese thought of concepts of magic correspondence and systematic correspondence, both of which are based on the principle that the phenomena of the visible and the invisible world stand in mutual dependence. Concepts like *yin*, *yang*, and *qi* are embedded in a system of correspondences that involve noncausal dependencies. For example, the movement of *qi* in the body is understood in part on the basis of the body having an upper part (*yang*) and a lower half (*yin*), and a left side (*yang*) and a right side (*yin*) (ibid., p. 88).

Thus traditional Chinese medicine is closer to prescientific assumptions of homeopathic magic, which employs the principle that like corresponds to like, than it is to modern conceptions of causality. Thagard (1988, ch. 9) describes how much prescientific and pseudoscientific thinking is based on resemblance rather than causality. The causal mode of explanation found in current scientific medicine has no room for explanations based on resemblance and mystical correspondences, so it is difficult to compare the two kinds of explanation head to head. Here the debate between traditional Chinese medicine and Western science has to move to a metalevel involving the efficacy of the different styles of explanation. Evaluating traditional Chinese herbal medicine is also very difficult from the perspective of Western, evidence-based medicine, because prescribed herbal remedies often involve mixtures of numerous kinds of herbs suggested by correspondence-based ideas. Determining the causal effect of a single herb would be viewed as pointless within traditional Chinese medicine.

Even here, however, there is not a complete breakdown of intelligibility, since traditional Chinese medicine does want to claim causal effectiveness for its treatments. Although it seems mysterious from the Western medical perspective why acupuncture places needles at certain points in the body that are thought to have the relevant correspondences, it is still possible to ask the question, common to both traditions, of whether the needling is causally effective. Hence the explanatory gap between traditional Chinese medicine and Western science is not so great as the gap, say, between Western science and fundamentalist religion, which claims that the primary source of evidence is a sacred text. Moreover, the gap between Western and Chinese medicine has shrunk over the centuries, in that the current explanatory role of systematic correspondences in Chinese medicine is much smaller than it was originally.

In sum, our discussion of the linguistic, conceptual, ontological, and explanatory differences between traditional Chinese medicine and traditional Western medicine has shown that the two approaches are not strongly incommensurable. Considerable mutual comprehension is possible, although it does not go so far as to permit translation of one conceptual system into the other; hence the two stems are weakly incommensurable. Let us now see how weak incommensurability affects the evaluation of acupuncture.

Evaluating Acupuncture

In the previous section, we saw that the linguistic, conceptual, referential, and explanatory differences between traditional Chinese medicine and Western medicine do not constitute insuperable barriers to their rational comparison, although the explanatory differences are more serious impediments. A Western researcher demanding that acupuncture and other therapeutic practices be evaluated with respect to their explanatory coherence as shown by randomized and blinded clinical trials would be stymied by a proponent of traditional Chinese medicine who said that all this was simply irrelevant. But traditional Chinese medicine is not a mystical religion; it is aimed at improving people's health, and its practitioners sincerely believe that it succeeds. Hence, even for the most orthodox practitioners of traditional Chinese medicine, there is an empirical standard, not just a doctrinal one.

In a head-to-head clash between Western and traditional Chinese medicine, it would be necessary to choose one of the conceptual-explanatory systems as superior and reject the other. A skeptical Western physician, for example, could argue that Western medicine has incontrovertible successes and that the whole Chinese system can be dispensed with. There is no reason, however, why evaluation of traditional Chinese medicine needs to be this absolute. Some prescientific medical practices, such as the North American aboriginals chewing salicin-containing willow bark to relieve pain, have turned out to be medically effective even by modern standards. It is entirely possible, therefore, that some traditional Chinese therapies such as acupuncture and herbal remedies might have some efficacy.

Acupuncture is a family of procedures, the most familiar of which involves penetration of specific points on the skin by thin metallic needles. If acupuncture were only comprehensible within traditional Chinese medicine, then it might indeed be concluded that acupuncture is strongly incommensurable with the substantially different system of Western medicine. But acupuncture has in fact been evaluated from the perspective of Western medicine, most recently and publicly by the NIH Consensus Development Conference that took place in November 1997. The operations of this conference are a striking example of evaluation occurring in the face of conceptual difficulties. Acupuncture would never have been invented within Western medical science, but that does not make it immune to scientific evaluation.

Like previous NIH Consensus Conferences, the acupuncture conference consisted in one and a half days of presentations followed the next morning by a presentation of a consensus report. This report was prepared by a twelve-member panel drawn from different backgrounds, including both acupuncture specialists and Western-trained medical experts (NIH, 1997). Panel members worked until 4 AM on the final day of the conference to reach agreement on a statement that was publicly released later that morning. The panel concluded that there is clear evidence that needle acupuncture is efficacious for adult postoperative and chemotherapy nausea and vomiting and probably for the nausea of pregnancy. It also found some evidence of efficacy for postoperative dental pain, and suggestive but not conclusive evidence for pain relief in other conditions such as menstrual cramps. Since acupuncture has minimal adverse effects, the panel stated that acupuncture may be a reasonable option for a number of clinical conditions such as stroke rehabilitation and osteoarthritis.

The panel reached its conclusions using the standards of Western medicine. Ideally, evaluation of medical effectiveness should be based on randomized, controlled, blinded clinical trials, but such trials have only been a part of medical research since World War II, and most Western medical practices are based on medical experience rather than rigorous tests. With a procedure as obvious as acupuncture, it is not easy to perform properly controlled experiments: unlike placebo pills, patients clearly know whether they have received acupuncture or not. Experiments using "sham" acupuncture, in which needles are inserted at nonstandard acupuncture points, have provided mixed results often intermediate between orthodox acupuncture and nontreatment. The panel decided not to insist on the highest standards of medical efficacy based only on rigorously controlled experiments, but rather to evaluate acupuncture based on the more usual clinical standards of Western medicine. The panel concluded that acupuncture may well be effective for the treatment of nausea and pain, and recommended future high-quality, randomized, controlled clinical trials on its effects.

The panel's recommendations were based on a large body of printed information provided by NIH in advance of the conference, and on the presentations of twenty-four speakers on the first day and a half of the conference. Although a few of the talks presented acupuncture within the context of traditional Chinese medicine, the vast majority discussed its effectiveness from the Western evidential perspective. Several talks discussed the possible neurochemical basis of acupuncture, presenting evidence that acupuncture stimulates the production of endogenous opioids and affects the secretion of neurotransmitters and neurohormones. The panel report, however, remained open to the traditional Chinese medicine based on *qi*: "Although biochemical and physiologic studies have provided insight into some of the biologic effects of acupuncture, acupuncture practice is based on a very different model of energy balance. This theory may provide new insights to medical research that may further elucidate the basis for acupuncture" (NIH 1997). This statement is not an endorsement of traditional Chinese medicine, but it suggests that its theory as well as its practice may turn out to be useful in Western scientific medicine. The implication, however, is that the theory of *qi* would need to be evaluated according to scientific standards, not in accord with traditional Chinese texts or the doctrine of correspondences.

According to some sociologists, science is essentially a power play in which some researchers marshal resources to triumph over others (Latour & Woolgar, 1986). One interpretation of the NIH consensus conference would be that acupuncture proponents managed to dominate by assembling speakers and panel members to endorse their claims. Alternatively, the conference organizers could conceivably have assembled a panel of hard-line Western medical researchers who would have dismissed acupuncture as pseudoscientific trickery. Although consensus conferences undoubtedly have a political dimension, their operation is designed to encourage evidence evaluation rather than political manipulation. The twelve panel members were presented with a common body of information to evaluate, and most of them had no strong interest for or against acupuncture; only two of the twelve were practicing acupuncturists, and both had Western medical training. Some of the studies they looked at, particularly the well-done and replicated studies concerning the effects of acupuncture on postoperative nausea, were very impressive.

Conceptual Change as Intentional and Emotional

We have argued that the substantial conceptual differences between traditional Chinese medicine and Western medicine can be overcome, but it would be rash to exaggerate the ease with which mutual understanding can be accomplished. Consider two people, one an expert on and a proponent of traditional Chinese medicine—C—and the other trained in Western medicine—W. Initially, C and W will scarcely be able to talk to each other, with the former using concepts like *qi* and the other using concepts like *germ* and *immune system*. Any degree of mutual comprehension that develops will depend not only on casual communication, which will be ineffective, but on the kind of *intentional* conceptual change that is discussed by Sinatra and Pintrich (2003). C and W each must have the motivation to acquire enough of the other's conceptual system that comparison and evaluation becomes possible. Only then will C and W have the capability of changing their conceptual systems by adopting components of the alternative system and by revising their own concepts (see Thagard, 1992, for a taxonomy of conceptual changes).

People who undergo conceptual change, whether from the traditional Chinese system of medicine to the Western system or vice versa, must have a set of cognitive goals that directs their thinking. First, they must have the goal of understanding the alternative system, which requires becoming familiar with (but not necessarily endorsing) its concepts, hypotheses, and evidence. Accomplishing this goal may involve trying to translate the alternative system into more familiar terms, or understanding the system on its own terms. Second, they must have the goal of assessing the alternative systems with respect to explanatory coherence and practical efficacy. Third, they must be willing to recognize an alternative conceptual system as superior in important respects to their own and therefore worthy of replacing it, partially or totally. Thus the development of mutual understanding and the process of conceptual change depend in part on the intentions of people to take seriously conceptual systems that differ from the ones they currently hold.

The goal of understanding an alternative system may be most effectively accomplished by intentionally striving to appreciate the kinds of linguistic, conceptual, referential, and explanatory differences discussed above. Learning a second language is a challenging task that requires much motivation. Becoming aware of conceptual differences in organization of kind hierarchies can also benefit from having the goal of noticing such differences. Differences in reference and explanatory style are often so subtle that they will not be noticed unless a thinker has the explicit goal of seeing how an alternative system makes claims about the world and explains occurrences in it.

One major impediment to conceptual change that has been largely ignored in psychological and philosophical discussions is the emotional attachment that people have to their own systems. Like all thinking, scientific cognition is in part an emotional process (Thagard, 2006a, ch. 10). People not only hold and use their concepts and hypotheses, they also feel emotionally attached to them and respond with negative emotions to concepts and hypotheses that clash with them. For a proponent of traditional Chinese medicine, acupuncture may be a revered practice associated with happy outcomes, whereas for a Western physician it may seem like a ridiculous throwback to prescientific practices held in contempt. Conceptual change about different approaches to medicine involves changing not only concepts, hypotheses, and practices but also emotional attitudes toward those concepts, hypotheses, and practices. Having the intention to understand and evaluate alternative views can make the emotional component of conceptual change more easily realized.

Emotional conceptual change can be understood in terms of alteration in the positive or negative *valence* attached to a concept or other representation. Concepts such as *baby* and *ice cream* have positive valence for most people, while concepts such as *death* and *disease* have negative valence. Emotional conceptual change is a change of valence from positive to negative or vice versa. Consider, for example, religious fundamentalists who encounter the Darwinian concept of evolution. Initially, this concept has negative valence because of its associations with scientific views that the fundamentalists see in conflict with their religious views, which have strong positive valence. For fundamentalists to accept the theory of evolution, they not only need to change their beliefs and concepts such as *human*, which is a kind of animal in the Darwinian scheme but not in the religious one, they also need to change the valence they attach to concepts such as *evolution*. Conversely, biologists who attach positive valence to the concept of evolution because of its role in a powerful and successful scientific theory would require emotional conceptual change if they were to become fundamentalists and view evolution as not only false but otiose. Similarly, proponents of Western and traditional Chinese medicine start with different emotional valences for concepts such as *germ* and *qi*, and adoption of the alternative approach to medicine will involve changes to these valences as well as cognitive changes to beliefs and concepts.

What are the mental mechanisms for emotional conceptual change? First, a concept can acquire a valence through emotionally charged associations, for example, a positive experience with ice cream or a negative one with sickness. Second, once a concept has a valence, new experiences can lead to different associations, as in someone who formerly disliked ice cream but grew to enjoy it. Third, and most dramatically, emotional conceptual change can be the result of emotional coherence, in which a whole network of representations is adjusted for both the acceptability and the emotional valences of the representations; see Thagard (2000) for an account of how emotional coherence can be computed. A dramatic shift in emotional coherence would be required for a fundamentalist to become a Darwinian or vice versa, and for an advocate of Western medicine to become a practitioner of traditional Chinese medicine or vice versa. Entrenched emotional attitudes may be a substantial barrier to such large-scale cognitive/emotional shifts.

This section has discussed conceptual change at a more individual level than earlier ones, which compared whole communities of medical practitioners. But the issues are essentially the same. For a community to undergo conceptual change, the individuals in it must undergo conceptual change. Such change is a social process as well as an individual one, since it often requires interactions between individuals. For a discussion of the need to integrate social and psychological explanations of scientific change, see Thagard (1999).

Conclusion

We embarked on this study in order to examine a more extreme case of possible incommensurability than typically occurs in the history of Western science. The issue is important because questions about incommensurability raised by Kuhn and Feyerabend are often used to support relativist views that challenge the rationality of science (Laudan, 1990). Our examination has shown that there are indeed linguistic, conceptual, referential, and explanatory differences that make mutual evaluation of traditional Chinese medicine and Western scientific medicine difficult. We have also seen, however, that these difficulties can to a great extent be overcome by earnest, intentional attempts to learn alternative languages, conceptual schemes, and explanatory patterns. As the NIH consensus conferences shows, a therapeutic practice like acupuncture can be evaluated for its effectiveness without adopting the theoretical framework from which it arose. We do not need to have a grand, holistic clash of traditional Chinese medicine versus Western scientific medicine to conduct a useful piecemeal evaluation of particular treatments. The two systems of medicine are weakly incommensurable (mutually untranslatable), but they are not strongly incommensurable (mutually unintelligible). Despite the substantial barriers to complete translation that divide different systems of medicine, rational scientific evaluation of practices such as acupuncture is possible. But such evaluation requires earnest intentions to understand alternative systems sufficiently to make comparison with more familiar methods possible.

For future research, much more needs to be done concerning the nature of emotional conceptual change. Does it involve more complex emotional changes than simply a shift of positive and negative valences, for example, ones involving full-fledged emotions such as love and disgust? Are the processes of emotional conceptual change that operate in science the same as ones that operate in other areas of life, for example, in psychotherapy? To what extent can emotional conceptual change be intentionally controlled? How frequently are emotions impediments to evidence-based cognitive change, and what does it take to overcome such impediments? Are nonverbal representations such as visual images also subject to emotional change? Answers to such questions should contribute to a deeper understanding of conceptual change in medicine, science, and everyday life.

16   Conceptual Change in Medicine: Explanations of Mental Illness from Demons to Epigenetics

Paul Thagard and Scott Findlay

Introduction

Many people are afflicted by mental illnesses such as depression and schizophrenia, but there has been much controversy about the nature and origin of such disorders. Are they the result of demonic possession, bodily imbalances, faulty childhoods, social control, genetic defects, nutritional inadequacies, or other factors? Are mental illnesses biological disorders like cancer and influenza, or are they mere social constructions? In the more than two millennia that people have thought about mental illnesses, many changes have come about in the conceptualization of psychological disease in general and of particular mental problems such as epilepsy. This chapter will describe the nature of these conceptual changes and provide a discussion of the current evolving state of deliberations about mental illness.

First we will provide a quick review of how explanations of classifications of mental illnesses have changed dramatically over the centuries. This review will show that the history of understanding of medical illnesses displays the radical kinds of conceptual change that have occurred in the history of science, including not only reclassification of diseases but also fundamental changes in the way that classification has been performed. Then we will argue that classification, diagnosis, and treatment of mental illnesses need to be based on multilevel explanations that take into account causal mechanisms that operate on at least four different levels: social, psychological, neural, and molecular. We will illustrate the complex causality of mental illnesses by describing recent research on epigenetics, which concerns the molecular control of gene activity by environmental factors.

Conceptual Change

In *Conceptual Revolutions*, Thagard (1992) provided a comprehensive account of kinds of conceptual change in major scientific revolutions, from Copernicus to Darwin to Einstein. Some conceptual change involves only minor modifications to existing concepts, as scientific research shows that a disorder such as schizophrenia can be treated by a newly developed drug. But more important modifications to concepts include the following:

• Introduction of new concepts such as *schizophrenia*.

• Abandonment of old ones such as *demonic possession*.

• Differentiation of concepts into subtypes such as *depression* to include *manic depression*.

• Reclassification of concepts from one category to another, for example, identifying epilepsy as a kind of bodily disorder instead of as a kind of divine visitation.

• Changing the ways in which classifications are done, for example, using underlying physical causes rather than observable symptoms.

We can find instances of all these kinds of conceptual changes in the history of thinking about mental illness.

Historians have identified many important developments in the description and explanation of mental illness (e.g., Porter, 2002; Shorter, 1997). The first important shift took place in Greece in the fifth century BC, when Hippocrates argued against the prevailing view that epilepsy was a "sacred disease" caused by visitations from the gods. Instead, in keeping with his general theory that diseases are caused by imbalances in the four humors (phlegm, blood, yellow bile, black bile, blood), Hippocrates argued that madness results from problems with the brain:

Men ought to know that from nothing else but the brain come joys, delights, laughter and sports, and sorrows, griefs, despondency, and lamentations. And by this, in an especial manner, we acquire wisdom and knowledge, and see and hear, and know what are foul and what are fair, what are bad and what are good, what are sweet, and what unsavory; some we discriminate by habit, and some we perceive by their utility. By this we distinguish objects of relish and disrelish, according to the seasons; and the same things do not always please us. And by the same organ we become mad and delirious, and fears and terrors assail us, some by night, and some by day, and dreams and untimely wanderings, and cares that are not suitable, and ignorance of present circumstances, desuetude, and unskilfulness. All these things we endure from the brain, when it is not healthy, but is more hot, more cold, more moist, or more dry than natural, or when it suffers any other preternatural and unusual affection. And we become mad from its humidity. For when it is more moist than natural, it is necessarily put into motion, and the affection being moved, neither the sight nor hearing can be at rest, and the tongue speaks in accordance with the sight and hearing. (Hippocrates, 400 BC)

The humoral theory of disease dominated European medicine for thousands of years until Pasteur's germ theory came along in the mid-nineteenth century. Mental illnesses such as epilepsy may not have anything to do with the moisture level of the brain, but Hippocrates made a great advance in recognizing them as problems of the brain rather than as the result of heavenly interventions, as the Assyrians, Egyptians, and earlier Greeks had assumed. Hence Hippocratic theories introduced a dramatic reclassification of mental illnesses as biological disorders rather than supernatural afflictions.

That reclassification was far from universally accepted, and theological and mystical explanations of mental illnesses survive even today. More sophisticated biological explanations were slow in coming, with few developments until the seventeenth century when physicians such as Thomas Willis, Archibald Pitcairn, and Herman Boerhave began to develop mechanistic ideas based on organs, nerves, fibers, and fluids that could be applied to madness. In the eighteenth century, explanations of mental functioning took a psychological turn through the writings of philosophers such as Locke and Condillac. During the nineteenth century, psychiatry developed as a branch of medicine in both France and Germany, with the French emphasizing psychological descriptions and explanations, and the Germans emphasizing the neurological. These developments generated a tension that survives today, between biological and psychological explanations of mental functioning. The next section will suggest a way of overcoming this tension by considering multilevel mechanisms.

The twentieth-century rise of Freudian psychoanalysis shifted psychiatry toward concern with psychological explanations. Although Freud began with a strong interest in neural mechanisms, he quickly moved on to psychological matters such as the workings of the unconscious and the lingering effects of childhood sexuality. In the second half of the twentieth century, psychiatry shifted back toward more biological directions. The major impetus in this direction came from the discovery of antipsychotic medicines in the 1950s that provided effective ways of reducing the symptoms of schizophrenia. Other important discoveries were the use of lithium for treating manic depression (also known as bipolar disorder) and the development of drugs for anxiety and depression. Initially, these drugs were used without much understanding of why they worked, but developments in molecular biology and neuroscience generated increasingly deep understanding of the underlying mechanisms by which drugs relieved the symptoms of many important mental illnesses through effects on such key neurotransmitters as dopamine and serotonin (Meyer & Quenzer, 2005; Thagard, forthcoming-c). Dramatic developments in genetics also contributed to understanding of the biological bases of mental illnesses, which epidemiological evidence showed to have strong (but not exclusively) inherited components.

It should be evident from our quick sketch of the conceptual developments in psychiatry that this history cannot be viewed as a simple accumulative process in which better ideas replace their inferior predecessors. The alternating swings between psychological and neurological views of mental illness involve noncumulative conceptual shifts, from thinking of mental illnesses as disorders with psychological causes to thinking of them as neurological, and back again. The crucial question is what kinds of condition mental illnesses are, with very different general and particular concepts arising from considering them in either psychological or neural terms.

In the 1960s, a much more radical reclassification was suggested by the "antipsychiatry" movement that denied the objectivity of both psychological and neurological accounts of mental illness. Writers such as Thomas Szasz (1961), R. D. Laing (1967), and Michel Foucault (1965) contended that mental illnesses are not objective disorders but rather social categories used to control unruly people. If this view had become dominant—and it still has supporters in some postmodernist circles—then mental illnesses ought to be reclassified as bogus social constructions. For reasons to resist such reclassification, see, for example, Thagard (2008).

Within mainstream psychiatry, mammoth conceptual development took place through the issuing of successive editions of the American Psychiatric Association's *Diagnostic and Statistical Manual* (DSM) in 1952, 1968, 1980, 1994, and 2000. The most striking change is the proliferation of disorders: the most recent version, DSM-IV-TR, has more than 900 pages and several hundred disorders. Not only have new concepts applying to distinct mental disorders been introduced, but some concepts were dropped in subsequent editions. For example, homosexuality was classed as a disorder in the first two editions, but not in the third and later ones. For historical and philosophical discussions of the DSM, see Kutchins and Kirk (1997) and Murphy (2006). DSM-V is now under construction with an attempt to make the resulting classifications more consistent with the increasing number of neurological findings about mental illness deriving from brain scanning and other technological developments.

This survey of the history of psychiatry has been very brief, but it provides evidence for the claim made at the beginning of this section that investigations of mental illnesses have introduced major conceptual changes. Attempts to explain and treat mental illnesses have led to classifications that at sometimes introduce new concepts, abandon old ones, and shift mental illnesses among categories. Controversies still reign about whether mental illnesses should be classified psychologically based on behaviors, neurologically based on underlying biological conditions, or sociologically based on social determinants. The next section will attempt to reconcile some of these competing viewpoints.

Multilevel Explanations in Medicine

Dictionaries often define an illness or disease as an impairment of normal physiological function. We want to propose a much broader conception of disease as a serious malfunctioning in a system of multilevel interacting mechanisms. This approach draws on recent work on mechanistic explanation in psychology, neuroscience, and other fields (see Thagard, 2006c, 2010a; Thagard, forthcoming-d; Bunge, 2003; Bechtel, 2008).

Modifying Bunge (2003), we can define a system as a quadruple, <Environment, Parts, Interconnections, and Changes>, EPIC for short. The parts are the objects (entities) that compose the system. To take a biological example, a body is composed of such parts as the head, torso, arms, and legs. The environment is the collection of items that act on the parts, which for a body would include other people and things that interact with it. The interconnections are the relations among the parts, especially the bonds that tie them together. In a body, key relations include the physical connections between head, neck, torso, and limbs. Finally, the changes are the processes that make the system behave as it does.

A person cannot be easily decomposed into a single EPIC system. Even an organ can be understood at multiple physical levels; for example, the human heart can be divided into various parts such as the chambers and valves, each of which consist of molecules, which consist of atoms, which consist of subatomic particles, which may consist of quarks or multidimensional strings. To characterize multilevel systems, we can generalize the EPIC idea and think of a multilevel system as consisting of a series of quadruples, with the structure:

$<E_1, P_1, I_1, C_1>$

$<E_2, P_2, I_2, C_2>$

. . .

$<E_n, P_n, I_n, C_n>$.

At each level, there is a subsystem consisting of the relevant environment, parts, interconnections, and changes. See Thagard (forthcoming-d) for further analysis and an extended argument that the human self should be understood as a system of interacting mechanisms operating at four levels: social, psychological, neural, and molecular. Now we will try to show the implications of this interpretation for understanding mental illness.

To provide a concrete example, consider the discussion of suicide by Goodwin and Jamison (2007) in their standard text on manic-depressive illness. They report that suicide is far more common among people with this illness than among the general population and that "it is clear that suicide is caused by a potent combination of biological and psychosocial risk factors" (p. 255). Their discussion includes the following factors:

• Genetic and family transmission: Partial heritability of suicidal behavior is shown by studies of families, twins, and adoptees.

• Neurobiological factors: There are associations of suicide with serotonin hypofunction, alterations in norepinephrine function, and hyperactivity in the hypothalamic-pituitary-adrenal axis.

• Psychological traits and temperament: A passive sense of hopelessness is a chronic risk factor for suicide.

• Social factors: Stressors such as job losses, relationship breakups, and legal proceedings can precipitate suicide.

These factors suggest that understanding manic-depressive illness will require attention to mechanisms at four levels: the molecular, neural, psychological, and social. Each of these can be characterized in EPIC terms with respect to Environment, Parts, Interconnections, and Changes.

The relevance of the molecular level is most clearly shown by the partial genetic heritability of bipolar illness and the success of the most common treatment for the disorder, administration of lithium, which affects levels of dopamine, serotonin, and other neurotransmitters. The functioning of these neurotransmitters only makes sense in the context of the interactive firing activities of billions of neurons, so the neural level is clearly relevant also. Thus at the molecular and neural levels we have interactions between two kinds of parts: chemicals such as dopamine at the molecular level, and cells such as neurons and astrocytes at the neural level.

At the psychological level, the relevant parts are less clear, but it is common in cognitive science to think of mental processes as resulting from the computational interactions of various kinds of mental representations, which can include concepts, rules, images, and analogies (Smith & Kosslyn, 2007; Thagard, 2005a). It is unusual to think of the mind as having mental representations as parts, but the explanatory successes of contemporary psychological theory justify this shift away from common-sense ways of talking about the mind. At the social level, the parts are obviously people whose interactions can have strong effects on the thoughts and behaviors of people in general, not just of bipolar patients who are spurred to suicide by negative events in their jobs, romances, or legal situations.

We are now in a position to offer a broad view of the nature and causation of mental illness. For a relatively simple illness such as Huntington's disease, causality can be characterized at one level, the molecular, because the disease originates from a mutation in a specific gene that produces a specific protein crucial for brain functioning. But proper mental functioning more generally requires good performance at all levels, including the neural, psychological, and social, as previously discussed. Correlatively, mental illnesses such as depression involve malfunctioning at multiple levels, from the molecular (inadequate serotonin levels) to the neural (patterns of neural firing) to the psychological (obsessive negative thoughts) to the social (bad relationships that can be both causes and effects of depression). Hence mental illnesses involve breakdowns in mechanisms at multiple levels, eliminating the need to settle the ancient controversies over whether the causes of madness are psychological, biological, or social. All of these levels are relevant to explaining both successful functioning of human beings and the serious kinds of malfunctioning found in mental illness.

The case for multilevel explanations of mental illness can also be made by reference to other diseases. Schizophrenia, for example, clearly has molecular causes, as shown by the high level of heritability found through twin studies and the efficacy of dopamine antagonists in its treatment. On the other hand, social and psychological causation is also evident from the fact that schizophrenia is strikingly more common in children of immigrants (Bentall, 2004; Cooper, 2005). Similarly, multilevel causality of depression is shown by the evidence that the most successful treatments tend to be combined ones that operate both at the molecular/neural levels via antidepressant medications and at the social/psychological levels via psychotherapy. Multilevel causality should not be surprising, since it is also found in nonmental illnesses such as cancer and heart disease, which result from factors ranging from the molecular—genetic mutation, high cholesterol—to the social—second-hand smoke, job stress (Thagard, 1999).

To illustrate the growing importance of multilevel explanations of mental illness, we will review an important recent development in medical science: the rapidly growing field of epigenetics.

Epigenetics

Scientific Explanations in Epigenetics

Many diseases have been linked to genetic malfunctions, and some, such as cystic fibrosis and Huntington's, are the result of specific kinds of mutations. However, the operations of genes are turning out to be far more complicated than previously suspected, leading to a new class of epigenetic diseases. Epigenetic explanations are important since they provide a mechanistic account of how environmental and genetic factors can interact in biological processes such as development and the onset of disease. We will describe how epigenetics provides a new kind of disease explanation schema and hence produces a change in the concept of disease, including mental illness.

"Epigenetics" refers to that which is applied "*upon* the genes." The term was first included in scientific explanations concerning developmental biology by Conrad H. Waddington in 1942 (Richards, 2006). Today, epigenetics usually refers to changes in the genetic material that affect gene expression independent of any differences in the DNA sequence itself. Central to the concepts of epigenetics are mechanistic explanations of how various modifications made to the genetic material can affect gene expression. Today, the emerging field of epigenetics is introducing a new level of explanation into the explanatory framework used by scientists to explain phenomena ranging from early embryonic development to the etiology of widespread mental illnesses.

The simplest illustration of the importance of epigenetics presents itself in mammalian development. Different adult cell types that make up various tissues appear wildly different to us, and they are. However, skin cells, liver cells, and every other cell in our own bodies were all derived from a small collection of genetically identical embryonic stem cells. How can two genetically identical cells give rise to two completely different adult cells? How does a liver cell know to divide to give rise to new liver cells and not skin cells after thousands of divisions? Answers to these questions are provided by epigenetics. Since the entire genome is replicated for each cell division, the information necessary to specify cell type cannot be contained within the DNA sequence itself. Instead, different transient environmental signals during early development turn on some genes and silence others, thus modifying gene expression to achieve the appropriate adult cell type. This is accomplished by modifying DNA-containing chromatin, the genetic material, in ways that are both semistable and heritable. The two main mechanisms behind these modifications involve the post-translational modification of proteins found within chromatin known as histones, and the methylation of individual units of DNA. We will discuss these mechanisms next, summarizing Levenson and Sweatt (2005).

Epigenetic Modifications and Their Mechanisms

Histones are proteins that make up chromatin along with DNA. Both components are closely associated in chromatin, with about 147 base pairs (structural units) of DNA wrapping around a core histone octamer (made of two copies of each of the histone proteins H2a, H2b, H3, and H4) to form what is called a nucleosome (reviewed in Tsankova et al., 2007). This association is mediated by the attractive force between the negatively charged phosphate groups that line the backbone of DNA, and the positively charged basic amino acid residues that are abundant in histones. The wrapping of DNA around histones provides one level of folding for the genetic material. Additional levels of folding occur for chromatin to be bundled into the chromosomes we can see under the microscope in our actively dividing cells.

Residues on slim "tails" that extend away from each of the core histone proteins can be modified by the addition of small chemical groups to affect the attraction between histones and DNA. Such modifications include acetylation, phosphorylation, methylation, ubiquitination, and sumoylation.

Acetylation of lysine (K) residues by a class of enzymes known as histone acetyl transferases (HAT) is associated with transcriptional activation. The positively charged lysine side chain is neutralized upon addition of an acetyl group. This weakens the interaction between histones and DNA and allows increased access of cellular machinery for transcription (making new RNA based on a DNA template) that will later be used to make protein. A long stretch of transcriptionally active chromatin can be referred to as euchromatin.

Conversely, histone deacetylases (HDAC) can act to remove an acetyl group, promote a stronger interaction between histone and DNA, and reduce or repress transcription. A long stretch of transcriptionally repressed chromatin can be referred to as heterochromatin.

Phosphorylation of serine residues found on histone tails is also associated with transcriptional activation and is catalyzed by protein kinase enzymes and reversed by phophatase enzymes.

In a similar fashion, histone methyl transferase (HMT) and histone demethylase (HDM) enzymes can catalyze the methylation and demethylation of lysine residues, respectively. However, methylation can be associated with either increased or decreased levels of expression, depending on the residues affected. The "histone code hypothesis" suggests that certain modifications made to certain histone residues are either associated with transcriptional activation or repression, and their combined effects determine the overall accessibility of chromatin for transcription in a given region.

In addition to histone modifications, special substitution of histone octamer components, the sliding of nucleosomes along DNA, and the outright eviction of nucleosomes all contribute to changes in gene expression. Collectively these mechanisms are referred to as chromatin remodeling.

DNA methylation is another common epigenetic modification. In mammals such as humans, methylation involves the covalent addition of a methyl group (CH_3) to the fifth position of cytosine (C) bases found in DNA. A cytosine base "tagged" in this fashion is referred to as 5-methyl cytosine (5mC). Enzymes that carry out this methylation are known as DNA methyl transferases (DNMT). Importantly, these enzymes only recognize a cytosine base as a candidate for methylation when it is found directly before a guanine (G) base. The two bases are only separated by a phosphate group and are commonly represented by "CpG." Areas where CpG sequences are clustered closely together and are found more frequently than would be expected by chance are called "CpG islands." These islands are common in the upstream regulatory regions of many genes. Hypermethylation of these regions is usually associated with transcriptional silencing of the downstream gene. Conversely, hypomethylation is usually associated with an active transcriptional state of the downstream gene. Methylated bases are also known to interact with other proteins involved in chromatin remodeling and histone modification, demonstrating an important link between these two epigenetic mechanisms.

Epigenetic Systems

The two mechanisms discussed above, as well as a third mechanism involving RNA silencing, contribute to several systems of epigenetic regulation in addition to the simple regulation of individual gene expression described above. Two of these epigenetic systems are X-chromosome inactivation and genetic imprinting (see Richards, 2006, for discussion). The former occurs in XX females since they have double the amount of X chromosome genes as their male (XY) counterparts and therefore achieve "dosage compensation" by silencing one entire X chromosome in every cell. Several RNA molecules play a major role by physically interacting with the X chromosome that is to be silenced (reviewed in Richards, 2006).

Genetic imprinting involves the hemizygous (from only one allele) expression of certain genes in a parent of origin-dependent manner. It is believed that around 100 to 200 human genes (http://igc.otago.ac.nz/home.html) are expressed from the paternal chromosome (inherited from the father) and remain silenced on the corresponding maternal chromosome (inherited from the mother) or vice versa. Many of the epigenetic mechanisms discussed above are involved in imprinting. It is important to

Normal State

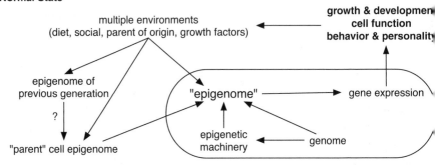

Figure 16.1
A simplified illustration of the framework of epigenetic explanations of biological phenomena. Solid arrows represent causal relations. The large oval indicates a single cell (e.g., a neuron) of interest. The question mark indicates that the possibility of transgenerational inheritance of epigenetic states in humans is controversial and still largely speculative.

understand that the effect of genetic imprinting is not that paternal and maternal chromosomes have a different sequence of DNA in a given region, but that expression of one parent's gene is encouraged while expression of the corresponding gene is silenced.

Epigenetics Can Explain Gene-Environment Interactions

Epigenetic explanations are very powerful since they provide a detailed mechanistic account for how both environmental and genetic factors might interact to affect biological processes such as development, behavior, and the onset of disease. In this framework, the effects of the environment and genetic factors are no longer mutually exclusive. Figure 16.1 provides a simplified illustration of how the epigenetic state of the entire genome known as the "epigenome" is influenced by genetic, epigenetic, and environmental factors.

Metastable Epialleles

The concept of a metastable epiallele is also important in epigenetic explanations. It can be defined as a region of the genome that can account for phenotypic differences between genetically identical cells (and individuals) based on epigenetic modifications that are both variable and reversible (Jirtle & Skinner, 2007). The best example of one such gene is the mouse

A^{vy} (viable yellow agouti) created by the insertion of an intracisternal A particle (IAP) transposable element upstream of the normal agouti (A) allele. In this transgenic system, the methylation state of the IAP determines if proper expression of the *agouti* gene takes place. Ectopic expression of the gene occurs when the IAP element is hypomethylated and is marked phenotypically by a yellow coat color, obesity, and diabetes. Highly methylated IAP induces normal expression of the *agouti* gene and results in a brown coat color. The degree of IAP methylation can vary in a random fashion, but interestingly can also be influenced by maternal diet during early development of offspring. Maternal diets supplemented with the methyl donor folic acid were shown to result in a higher proportion of brown pups. This effect was indeed mediated by increased methylation of CpG sites in the IAP (Wolff et al., 1998; Waterland & Jirtle, 2003; Cooney, Dave & Wolff, 2002). Although a transgenic model was used, this is a great illustration of how one small environmental component can have great effects on gene expression as mediated by epigenetic and genetic mechanisms. These types of explanations may prove to be central not only to growth and development, but to the etiology of human conditions as complex as cancer, and mental illnesses such as major depressive disorder (MDD).

Epigenetic Explanations of Disease

Thagard (1999) uses a collection of *explanation schemas* to outline the causes of numerous diseases. Today, our knowledge calls for the addition of an epigenetic explanation schema to such explanations of disease. Epigenetics might be able to help explain aspects of complex multifactorial diseases for which good explanations do not always exist. An explanation schema for diseases classified as epigenetic would look something like this:

Explanation target:

Why does a **patient** have a **disease** with associated **symptoms**?

Explanatory pattern:

The **patient** has an **epimutation** affecting the accessibility of **gene(s)** in a given region of **chromatin**.

Improper expression of **gene(s)** produces the **disease** and its **symptoms**.

"Epimutation" is used here to mean "any mistake in the epigenetic modification(s) of a particular region of chromatin." The concept is based on that of "mutation," which refers to a mistake in the sequence of DNA itself. An epimutation can originate in the affected cell or can be inherited upon division of its "parent cell." The ultimate cause of an epimutation in either case is either genetic or environmental in nature.

"Environment" is used in a broad sense to account for a wide range of nongenetic factors that might be implicated in the etiology of complex disease. At the molecular level, many such factors have been shown to affect the epigenome. For example, a small region of chromatin might have one pattern of DNA methylation or another depending on if it was inherited maternally or paternally. Exposure to certain chemicals during pregnancy can affect the state of an offspring's epigenome (reviewed in Szyf, 2009). DNA methylation patterns can change in response to nutritional restriction during gestation in rats (MacLennan et al., 2004; Ke et al., 2006; Sinclair et al., 2007; all reviewed in Szyf, 2009). Additionally, a maternal diet deficient in folic acid has been shown to have a wide variety of teratogenic effects, perhaps due to reduced availability of methyl groups usually stripped from folic acid to ultimately be used for DNA methylation (Wainfan et al., 1989; Pogribny, Miller & James, 1997; Brunaud et al., 2003; all reviewed in Szyf, 2009). The early social environment can also affect behavior in a way that is mediated by epigenetics. For example, attenuated hypothalamic-pituitary-adrenal (HPA) stress response has been associated with high levels of glucocorticoid receptor (GR) mRNA (Jacobson & Sapolsky, 1991). Rat pups raised by mothers who licked and groomed their pups more often showed higher levels of GR mRNA in the hippocampus than those raised by mothers who licked and groomed their pups less. This effect was independent of genetic and prenatal factors (Champagne et al., 2003). Importantly, this increase in GR was associated with decreased HPA stress response as expected (Caldji, Diorio & Meaney, 2000). This association appears to be mediated by decreased methylation of the GR 1_7 gene promoter in the hippocampus that is established during the first week of life and retained into adulthood (Weaver et al., 2004). See Champagne and Curley (2009) for a review of stress response studies.

Genetic causes of epimutation are usually due to a simple mutation in a gene that normally codes for a component of epigenetic regulation such as a HAT or DNMT. A simplified illustration of the causes of epimutation and how they contribute to complex disease is shown in figure 16.2.

Possible Diseased State

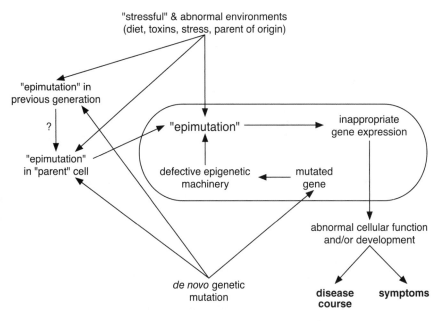

Figure 16.2
Illustration of a generalized diseased state due to epimutation. Solid arrows indicate causal relations. The large oval indicates a single cell (e.g., a neuron) of interest. The question mark indicates that the possibility of transgenerational inheritance of epigenetic states in humans is controversial and still largely speculative.

Types of Epigenetic Diseases

Like some other diseases, we can think about epigenetic diseases in terms of a hierarchy of schemas (figure 16.3) made up of the general epigenetic disease schema, different explanation schemas depending on the type of epigenetic modification or system that is being perturbed in each disease, and lastly, schemas for each individual disease (reviewed in Tsankova et al., 2007). It is important to note that these categorizations are very simplified, and some diseases arise because of the perturbation of multiple epigenetic mechanisms and other nonepigenetic factors.

Disorders with the most well-characterized epimutations are those that are concerned with the epigenetic system of genetic imprinting. For imprinted genes, the environmental input that affects gene expression is as simple as whether a given region of chromatin was inherited paternally

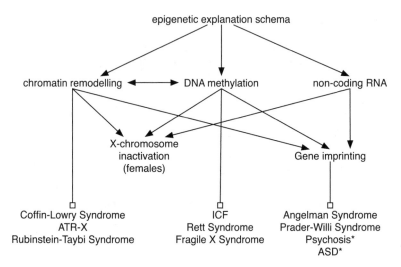

Figure 16.3
Hierarchy of schemas for epigenetic diseases. Solid arrows indicate causal relations. Square arrows indicate examples of mental disorders resulting from improper functioning of the connected epigenetic mechanism or system of regulation. ATR-X: Alpha-thalassemia/ mental retardation syndrome, X-linked. ICF: Immunodeficiency-centromeric instability-facial anomalies syndrome. ASD: autism spectrum disorders. An asterisk indicates disorders with a recently proposed role for epigenetics that has not been as well characterized. Most of these conditions are discussed by Tsankova et al. (2007).

or maternally. Since imprinted genes are normally expressed from only one chromosome, inheriting two chromosomes from one parent (uniparental disomy) or having a significant deletion of DNA in an imprinted region of only one chromosome can have drastic consequences. Such errors can result in certain imprinted genes being overexpressed or not expressed at all. Improper imprinting of a segment of chromatin at chromosome 15q11–13 has been implicated in two different neurodevelopmental disorders. Exclusively paternal (loss of maternal) expression in this region yields Angelman syndrome (AS) and exclusively maternal (loss of paternal) expression yields Prader-Willi syndrome (PWS). Symptoms of AS include cortical atrophy and cognitive abnormalities, while symptoms of PWS include mild mental retardation and endocrine abnormalities (Davies, Isles & Wilkinson, 2005). Crespi (2008) also proposes that alterations in genetic imprinting have played a central role in the evolution of a wide variety of mental illness. He proposes that imprinting errors favoring maternal expression of genes are associated with psychotic spectrum conditions, while imprinting errors favoring paternal expression of genes are associated with autism spectrum conditions.

Another well-understood disorder where normal functioning of an epigenetic modification is disrupted is Rett syndrome. This progressive neurodevelopmental disorder affects females when they reach 6–18 months of age and is characterized by symptoms such as deceleration of head growth, loss of speech and purposeful hand movements, mental retardation, and some autistic features (Chahrour & Zoghbi, 2007). It is known that mutations in the gene encoding methyl-CpG-binding protein 2 (MeCP2) are often the ultimate cause of Rett syndrome (Amir et al., 1999). MeCP2 normally functions to mediate interactions between the two epigenetic mechanisms of DNA methylation and chromatin remodeling. The protein has been shown to have a silencing effect on gene expression (Nikitina, Ghosh et al., 2007; Nikitina, Shi et al., 2007), but recent evidence suggests that MeCP2 likely also acts as an activator, and loss of transcriptional activation likely accounts for the majority of associated symptoms (discussed in Delcuve, Rastegar & Davie, 2009). Rett syndrome is a great example of an epigenetic disease with a clear genetic cause (mutation of MeCP2) as illustrated in the bottom portion of figure 16.2.

Epigenetics in Complex Mental Illnesses

In general, scientists use two basic criteria to determine if epigenetics might play a role in the etiology of complex disease: (1) an unexpected change or alteration in gene expression associated with the diseased state; and (2) evidence of one or more epigenetic mechanisms mediating this change. These criteria have been met for several complex mental illnesses such as autism spectrum disorders (ASDs), major depressive disorder (MDD), schizophrenia, and Alzheimer's. The first two of these will be discussed here.

The neurodevelopmental disorders mentioned above are categorically similar to ASD. LaSalle, Hogart, and Thatcher (2005) have proposed a "Rosetta Stone" approach to understanding possible epigenetic aspects of ASD based on less complex disorders that clearly involve epigenetics such as Rett syndrome. LaSalle has revealed that MeCP2 expression is often reduced in the frontal cortex of ASD patients despite any detectable genetic defects. This reduction of expression has been linked to abnormal methylation of the MeCP2 promoter (Nagarajan et al., 2006). Improper expression of MeCP2 has been shown to induce lower UBE3A and GABRB3 gene expression that likely contributes to the etiology of ASD (Samaco, Hogart & LaSalle, 2005). In addition to MeCP2 mutations as candidates to explain the etiology of ASD, the epigenetic system of genetic imprinting may also play a role, since between 2 percent and 42 percent of AS and PWS patients could also be diagnosed for ASD (LaSalle, Hogart & Thatcher, 2005). Future research is being conducted to determine how environmental agents such as bisphenol A (BPA) might contribute to abnormal methylation in mice and, together with defects in MeCP2 expression, might contribute to ASD.

Major depression is a serious disorder that is estimated to affect 16.2 percent of Americans at some point in their life (Murray & Lopez, 1997). After years of research looking for markers of genetic susceptibility there is still a great deal of uncertainty concerning etiological risk factors for depression and an absence of any detailed mechanisms. Mill and Petronis (2007) suggest that revealing a potential epigenetic layer of the etiology of MDD might account for several unexplained aspects of the complex disease. First, there exists a great deal of divergence between two genetically identical monozygotic (MZ) twins when it comes to MDD. While such differences have traditionally been attributed to their nonshared environment, there is evidence that epigenetic differences between MZ twins might accumulate over time in a random fashion or because of environmental factors (Fraga et al., 2005). In this way, epigenetics might account for a lack of convincing evidence as to the specific environmental risk factors for MDD, since differences between MZ twins would be accounted for by diverging epigenomes rather than the nonshared environment. Second, the possibility of transgenerational epigenetic inheritance may explain why it has been difficult to identify specific genes associated with MDD despite its high heritability, since it may not be specific sequences of DNA that confer the risk, but rather certain patterns of expression. Third, epigenetics would provide the mechanistic link needed to understand several reports of specific gene-environment interaction that might contribute to the development of MDD. Fourth, the finding that hormones have been shown to affect both chromatin remodeling (Csordas, Puschendorf & Grunicke, 1986; Pasqualini, Mercat & Giambiagi, 1989) and

DNA methylation (Yokomori, Moore & Negishi, 1995) might provide a way to explain the high prevalence of MDD among females. Fifth, genetic imprinting may be the mechanism of several parental origin effects reported for MDD.

Some of the best evidence in support of a role for epigenetics in both the onset and treatment of MDD comes from a mouse condition analogous to human depression known as chronic social defeat stress (Tsankova et al., 2007; Berton et al., 2006). In this model, exposure to an aggressor brings about lasting changes in the behavior of the mouse including social avoidance. While acute treatment with the antidepressant imipramine is insufficient to alleviate these symptoms, chronic treatment is effective in doing so (Berton et al., 2006, Tsankova et al., 2006). This is consistent with a slow response to antidepressants that act to increase levels of certain neurotransmitters in humans, and suggests that some downstream process must also be involved in MDD. In the mouse, chronic defeat stress causes decreased expression of two splice variants of brain-derived neurotrophic factor (*Bdnf*) in the hippocampus (Tsankova et al., 2006). This was mediated by an increase in H3-K27 dimethylation (a repressive modification) specifically at *Bdnf* promoters. This modification was long lasting and not reversed by antidepressant treatment. However, chronic treatment with imipramine induced H3 acetylation and H3-K4 methylation (activating modifications) at the same promoters. This was achieved by down-regulating *Hdac5* expression (Tsankova et al., 2006), and the same effect was achieved using the HDAC inhibitor sodium butyrate (Tsankova et al., 2006; Schroeder et al., 2007). This is a very clear case of epigenetic mechanisms mediating the onset and alleviation of symptoms in a mouse model of mental illness.

Further evidence of a role for epigenetics in the pathogenesis of MDD comes from postmortem samples of suicide victims. Expression of the epigenetic component DNMT was altered in several brain regions including the frontopolar cortex of suicide victims relative to individuals who died suddenly of other causes. This is evidence of altered gene expression associated with the suicide brain mediated by DNA methylation. The effects of altered DNMT expression included three CpG sites in the promoter region of a GABA receptor subunit gene that were found to be hypermethylated relative to control subjects causing decreased expression (Poulter et al., 2008).

Conclusion

We have gone into detail on the new epigenetic approach to explanations of mental illness for two reasons. First, this development illustrates the increasing complexity of multilevel explanations of psychiatric disorders. Social and molecular causes can interact in part because environments affect how genes are activated to produce or not to produce different proteins. Second, epigenetics appears to be introducing a new way of classifying diseases in terms of underlying complex causation and hence may be introducing an important kind of conceptual change in psychiatry. The development of scientific medicine in the nineteenth and twentieth centuries provided a very useful set of categories for causally classifying diseases as infectious, nutritional, autoimmune, and so on (Thagard, 1999). But some diseases such as arteriosclerosis defy such easy categorizations and can only be identified as multifactorial, a category that also now seems appropriate for most mental illnesses.

At the beginning of this chapter, we described some of the major conceptual changes that have taken place in the history of the scientific investigation of mental illness. These included the shift from religious to biological explanations, the rise and fall of psychoanalytic explanations, and major advances in the neurochemical basis of many disorders. We argued, however, that the latter advances do not obviate attention to psychological and social factors, rendering mental illnesses as best understood as multifactorial diseases. The new field of epigenetics provides dramatic details about how such causality, ranging from the social to the molecular, might work. Clearly, we should expect future conceptual changes in psychiatry.

V New Directions

Introduction

The cognitive science of science has largely concentrated on the nature of the concepts, theories, and explanations that scientists develop and students learn. There has been a general neglect of the *values* that are also part of scientific practice. The development of science has depended on valuing observations, experiments, and objectivity in the assessment of scientific hypotheses. The adoption of these values was a crucial part of the scientific revolution and the Enlightenment of the sixteenth through eighteenth centuries, and the values have remained important in the current practice of science. Today, however, they are challenged from various directions, ranging from the religious right who prefer faith over evidence, to the postmodernist left who decline to "privilege" science and rationality over obscurity and irony.

Whereas the role of values has largely been neglected in psychological and computational studies of science, philosophers of science have become increasingly concerned with issues about values (e.g., Douglas, 2009; Lacey, 1999; Longino, 1990). The major debate concerns whether there should be a role for social and ethical values in the assessment of scientific theories. The traditional view is that science should be "value free," which means that it is restricted to epistemic values such as evidence and truth, ignoring social values that might lead to biases in scientific research. A common worry is that allowing an incursion of social values into science will produce corruptions like those witnessed frequently in the history of science, for example, the racist distortions in Nazi science and the sexist distortions in nineteenth-century neuroscience. In response, philosophers such as Douglas (2009) argue that values inescapably and legitimately play a role in scientific deliberations, and the key to improving science is to make application of these values rationally appropriate as opposed to irrationally biased.

This chapter explores what cognitive science can contribute to this normative enterprise of finding an appropriate place for values in science by addressing questions that are largely ignored in philosophical discussions:

1. What are values?
2. How can values bias scientific deliberations?
3. How can values legitimately contribute to scientific inferences?

I will argue first that values are emotionally valenced mental representations, and will show how interconnected groups of values can be represented by cognitive-affective maps, a new technique for displaying concepts and their emotional connotations. Second, I will use cognitive-affective maps to show how values can distort scientific deliberations in ways that diminish rationality. Third, more positively, I will describe how values can legitimately affect scientific developments at all three stages of research: pursuit of discoveries, evaluation of results, and practical applications.

What Are Values?

Most scientists operate with epistemic (knowledge-related) values such as evidence and truth, and also with social values such as human welfare. We can contrast three opposed views of what values are: ideal structures, social constructions, and mental representations. On the first view, values are ideal structures such as Platonic forms or abstract linguistic structures. On the second view, values are mere complexes of behavior that result from social interactions. On the third view, which I want to defend here, values are mental representations with emotionally valenced meanings. Here "valence" is the term psychologists use to indicate emotionally positive associations (e.g., to concepts such as *baby*) or negative associations (e.g., to concepts such as *death*).

Consider, for example, the value *truth*. On the ideal structure view, truth is an abstract concept that exists independently of human minds, and the value attached to it is an abstract relation between a person (also an ideal entity) and the abstract concept. Philosophers often write of propositional attitudes, which are supposed to be relations between two kinds of abstract entities: persons and propositions, which are the meanings of sentences. Analogously, the ideal structure view of values can understand them as "conceptual attitudes," that is, relations between persons and the abstract meanings of concepts.

In contrast, the social construction understanding of truth views it as a way of talking that emerges from interactions among people. Shapin (1994) writes of the "social history of truth," taking truth to be a product of collective action rather than correspondence to reality. From this perspective, truth and other scientific values are just preferences that people happen to have as the result of their social circumstances.

Cognitive science offers an alternative to ideal structure and social construction accounts of the nature of values. As chapter 1 described, the standard assumption in cognitive science is that inferences such as decisions result from mental representations and processes. A mental representation is a structure or process in the mind that stands for something. Typical examples are concepts, which stand for classes of things; propositions, which stand for states of affairs; and rules, which stand for regularities in the world. In contrast to the ideal structure position, concepts and other mental representations are not abstract entities, but rather parts of the world as brain processes in individual humans (Thagard, 2010a). As chapter 18 will argue, the structure of a concept can be highly complex, with pointers to visual and emotional information.

The emotional component of concepts and rules shows how values can be mental representations. Consider, for example, the concept *car*. Brain representations of cars include verbal information such as that a car is a kind of vehicle and that there are many kinds of cars, for example, sedans and convertibles. As theorists such as Barsalou et al. (2003) have argued, the representation of the concept *car* also includes sensory information, such as images of the visual appearance of a car, as well as its sound and perhaps even its smell. For some people, cars may be emotionally neutral, but for many people the mental representation of a car will include either a positive or negative emotional association. For example, a race car driver probably thinks of cars very positively, whereas an environmentalist concerned with pollution may well think of them very negatively. In psychological terms, we can say that cars have a positive valence for the racer, but a negative valence for the environmentalist. Valences can be modeled computationally in various ways, including the attachment of a valence to the units in a localist neural network (ch. 5, this vol.; Thagard, 2006a), or as convolution between neural populations (ch. 8, this vol.).

Because part of the representation of a concept is an emotional valence, values can easily be understood as valenced representations. My valuing cars consists in having a mental representation of cars that includes a positive emotional valence arising from activity in brain areas relevant to assessing gains, such as the nucleus accumbens. My disvaluing of pollution consists in my having a mental representation that includes a negative emotional valence arising from activity in brain areas relevant to assessing losses, such as the amygdala. For a neurocomputational model of assessment of gains and losses, see Litt, Eliasmith, and Thagard (2008).

The claim that values are emotionally valenced mental representations does not entail that they are purely subjective, or just a matter of personal tastes. Concepts can vary in how well they represent the world: *tree*, *electron*, and *virus* are good examples of successful representations; *demon*, *phlogiston*, and *aether* are examples of representations that the march of science has discarded. Similarly, people can be wrong in their values, for example, mistakenly believing that cheeseburgers are good for people. In line with Thagard (2010a), I hold that positive values are objectively right if they promote the satisfaction of real human needs, including both basic biological ones and also psychological ones such as relatedness to other people.

Mapping Values

Assuming that individual values consist of emotionally valenced mental representations, we can now consider the nature of systems of them. Values do not occur in isolation: scientists typically have connections among values such as truth, evidence, and objectivity, whereas a religious thinker might prefer a different set of values such as God, faith, and obedience.

To display scientific value systems, we can use a technique that I developed for understanding conflict resolution (Thagard, 2010b, forthcoming-c; Findlay & Thagard, forthcoming). Cognitive-affective maps are graphical diagrams that show both cognitive structures (concepts, goals, beliefs) and their emotional valences; "affect" is the term used by psychologists and neuroscientists for matters related to emotion, mood, and motivation.

The following conventions are used in these maps. Elements (mental representations) are depicted by shapes:

- Ovals represent emotionally positive (pleasurable) elements.
- Hexagons represent emotionally negative (painful) elements.
- Rectangles represent elements that are neutral or carry both positive and negative aspects.

Straight lines depict relations between elements:

- Solid lines represent the relations between elements that are mutually supportive.
- Dashed lines represent the relations between elements that are incompatible with each other.

Moreover, the thickness of the lines in the shape represents the relative strength of the positive or negative value associated with it. Figure 17.1 illustrates the conventions used in representing values.

Figure 17.2 is a cognitive-affective map of my own set of epistemic values, indicating positive values such as truth and objectivity, as well as conflicting negative values such as error and bias. Other values not displayed include simplicity and analogy, both of which are part of my theory of explanatory coherence, which I take to be the main route to truth and objectivity (chs. 5, 7, this vol.). The structure in figure 17.2 is not just an abstract diagram, but also represents a neural network of the HOTCO sort (ch. 5, this vol.) that can be used as a computational model of why scientists make the decisions that they do. For example, if religious people advocate ignoring the lack of solid evidence for a miraculous event, then the structure displayed in figure 17.2 will reject their claims, because

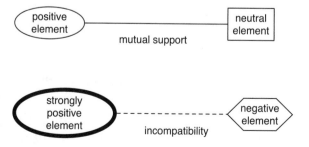

Figure 17.1
Conventions for cognitive-affective mapping.

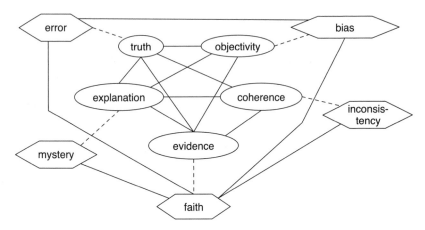

Figure 17.2
Epistemic values relevant to science. Concepts in ovals are positive values, whereas concepts in hexagons are negative values. Solid lines indicate mutual support, and dotted lines indicate incompatibility.

of the positive value attached to evidence and the negative value attached to faith.

Not every scientist or philosopher has the set of values shown in figure 17.2, but I view this group of values as defensible for reasons set out in many publications (e.g., Thagard, 1988, 1992, 1999, 2000). Others might choose to emphasize other values such as prediction, falsification, or empirical adequacy; I invite readers to draw their own cognitive-affective maps expressing their positive and negative values.

The same technique can be used to map social and ethical values, as shown in figure 17.3. This particular set of values is my own, reflecting an ethical view (needs-based consequentialism) defended elsewhere (Thagard, 2010a). These values differ from those of people who base their ethics on religion or a priori arguments. On my view, social values arise from the empirical fact that people universally have a set of biological and psychological needs. The biological needs are easiest to establish, as people obviously suffer severe harm if deprived of food, water, shelter, and health care. More contentious are the primary psychological needs of relatedness, competence, and autonomy, but there is evidence that these are also vitally important for human well-being (Deci & Ryan, 2002; Thagard, 2010a). Competence is the need to engage optimal challenges and experience physical and social mastery. Autonomy is the need to self-organize and

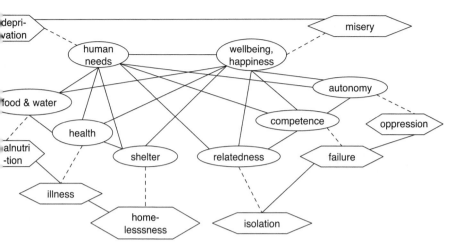

Figure 17.3
Social and ethical values related to satisfaction of human needs. Ovals indicate positive values, and hexagons indicate negative values.

regulate one's own behavior and avoid control by others. Relatedness is the need for social attachments and feelings of security, belongingness, and intimacy with others. Figure 17.3 shows the internal coherence among vital human needs and their incompatibility with a host of human conditions I disvalue, such as malnutrition and oppression.

Displaying sets of epistemic values (figure 17.2) and social values (figure 17.3) makes it possible to explore how they might interact in decisions concerning science. In practice, however, the decisions of particular scientists are not driven only by epistemic and social values, but also by personal goals such as success, fame, and prosperity. Figure 17.4 displays some of the personal values that can influence scientific decisions. This list is intended to be descriptive (what people often do value) rather than normative (what they ought to value). Normative assessment would require evaluation of the personal values shown in figure 17.4 with respect to the ethical values represented in figure 17.3. There are some obvious connections: wealth helps to satisfy the biological needs for food and shelter, and family helps to satisfy the psychological need for relatedness. The big difference, however, is that the social values in figure 17.3 are intended to take into account all the people affected by a decision, whereas the personal values in figure 17.4 operate just at the individual level.

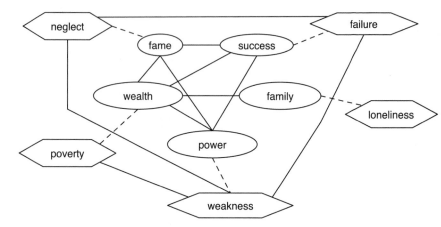

Figure 17.4
Personal values of some scientists. Values that are positive for a particular scientist are represented by ovals, and negative values are represented by hexagons. The different priority that different individuals place on these values could be shown by thickening the lines in the ovals and hexagons.

Appropriate Uses of Values in Science

I will now try to develop a general model of how the three kinds of values—epistemic, social, and personal—do and should interact. Then it will be possible to identify cases where science becomes biased because epistemic values are slighted in favor of social or personal values.

Different kinds of values are relevant to different contexts in scientific research and development. Philosophers of science commonly distinguish among the contexts of pursuit, evaluation, and application of scientific ideas. The context of pursuit includes deciding what research projects to pursue, setting goals, carrying out experiments, and generating new explanatory hypotheses. The context of pursuit includes not only discovery, but also all the planning that needs to go into figuring out where discoveries might be achieved. The context of evaluation (also called the "context of justification") assumes that evidence has been collected and hypotheses have been generated. According to the view presented in Part II, evaluation consists of inferences to the best explanation in which hypotheses are accepted or rejected depending on their explanatory coherence with respect to all the available evidence and competing hypotheses. Finally, the context of application includes decisions about how to use validated scientific ideas for practical purposes such as medicine, engineering, and social policy.

The context of pursuit legitimately mingles epistemic, social, and personal values. Scientists who are putting together a research plan can appropriately take into account *all* of the following questions concerning a potential project:

1. To what extent is the project likely to contribute to epistemic goals such as accumulating evidence and leading to the acceptance of explanatory theories?

2. To what extent is the project likely to contribute to human well-being as the result of satisfaction of human needs?

3. To what extent will the project further the professional careers of the scientists involved?

Different kinds of research will vary with respect to the importance of epistemic and social goals. To take examples from cognitive science, we can distinguish between basic research about fundamental psychological and neural processes, on the one hand, and applied research aimed at improving practices in education, mental health, and other areas relevant to human well-being, on the other.

Should scientists take into account their own professional success in deciding what research projects to pursue? There are at least two reasons for allowing personal values to be appropriate in the context of pursuit. First, to disallow them would be to insist that scientists be moral and epistemic saints, always ignoring their own interests in favor of general ethical and scientific goals. Few scientists are psychologically capable of either moral or epistemic sainthood, so a normative stance that requires perfection is unreasonable.

Second, eliminating the pursuit of personal interests in science may inhibit rather than enhance the accomplishment of the epistemic and social goals. Scientific work is often arduous and frustrating, and the personal motivations that come from dreams of personal success may be part of the motivation that keeps researchers toiling in the lab and stewing up new ideas. Moreover, the goal of personal success has the advantage of promoting diversity, encouraging researchers to undertake novel and risky research projects: there is more glory from work that is recognized as highly original.

The same considerations legitimize the pursuit of personal goals in the context of application. Turning ideas from physics, chemistry, biology, or cognitive science into socially valuable projects or commercial products naturally involves both epistemic values such as working with true theories and social values such as promoting well-being. But people applying science may also be driven by personal goals such as wealth, for example, technologists who apply scientific ideas to become billionaires. Without sinking to the egoistic view that only greed is good, we can still recognize that personal values can contribute to the overall success of applied projects because of their contributions to motivation, diversity, and novelty. The prospect of personal success can strongly encourage applied scientists such as technologists to work extraordinarily hard on risky projects that may have large payoffs, including ones that promote social well-being and may even stretch scientific theory in valuable directions. Of course, personal and social goals can also distort or interfere with epistemic ones, in the pursuit and application of science; the next section will consider psychological reasons why biases occur.

If the view I have been developing is correct, then decisions about the pursuit and application of science are multifactorial, involving consideration of the extent to which different projects promote the satisfaction of a full range of epistemic, social, and personal values. Descriptively and normatively, the process of decision is not reducible to some straightforward algorithm such as maximizing expected utility, but requires rather a complex process of constraint satisfaction that can include revising goals as well as choosing actions (Thagard, 2010a, ch. 6). Figure 17.5 shows a decision problem that is neutral between two projects, but which could be

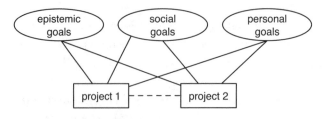

Figure 17.5
Map of decision between two competing projects, taking into account three kinds of goals. Ovals indicate positive values, and rectangles indicate neutrality. Solid lines indicate support, and the dotted line indicates incompatibility.

settled if there were stronger links of support between one project and the various kinds of goals.

I have argued that social and personal values legitimately contribute to scientific decisions in the contexts of pursuit and application, but the role of values in the context of evaluation is much more contentious. The claim that social and personal values should have *no* impact on evaluation and acceptance of theories has substantial appeal. Normative models for scientific inference (including explanatory coherence, Bayesian probability, and the confirmation theory of the logical positivists) all legislate that the evaluation of hypotheses should be done on the basis of evidence and reasoning alone, with no value component. We should accept theories that fit best with the evidence and reject the alternatives. In all these schemes, there is no room for the social and personal contributions to decision making shown in figure 17.5. Instead, the decision situation is limited to the sparse arrangement mapped in figure 17.6.

In contrast, Richard Rudner (1961, pp. 32–33) argued that social goals can legitimately play an indirect role in the assessment of hypotheses:

Since no scientific hypothesis is ever completely verified, in accepting a hypothesis on the basis of evidence, the scientist must make the decision that the evidence is sufficiently strong or that the probability is sufficiently high to warrant the acceptance of the hypothesis. Obviously, our decision with regard to the evidence and how strong is "strong enough" is going to be a function of the importance, in the typically ethical sense, of making a mistake in accepting or rejecting the hypothesis. Thus, to take a crude but easily manageable example, if the hypothesis under consideration stated that a toxic ingredient of a drug was not present in lethal quantity, then we would require a relatively high degree of confirmation or confidence before accepting the hypothesis—for the consequences of making a mistake here are exceedingly grave by our moral standards. In contrast, if our hypothesis stated that,

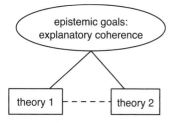

Figure 17.6
Theory acceptance as a decision problem concerning only epistemic goals.

on the basis of some sample, a certain lot of machine-stamped belt buckles was not defective, the degree of confidence we would require would be relatively lower. How sure we must be before we accept a hypothesis depends on how serious a mistake would be.

Here social goals do not contribute directly to the evaluation of the evidence for and against scientific theories, but do contribute indirectly by setting a threshold for acceptance, making it higher in cases where acceptance might have harmful consequences. Douglas (2009) defends at greater length a similar position. Neither Rudner nor Douglas proposes the opposite of this harm principle—a benefit principle that would allow hypotheses that would be beneficial if true to be accepted with a lower standard of evidence.

A similar issue arises in the law (Thagard, 2006a, ch. 9). In the British legal system, people accused of crimes are not supposed to be convicted unless the prosecution shows that they are guilty *beyond a reasonable doubt*. This constraint operates as well in the United States, Canada, and some other countries. In legal discussions, there is no accepted definition of what constitutes a reasonable doubt, but my preferred interpretation is shown in figure 17.7. In civil trials, decisions are based on *a preponderance of evidence*, which allows a hypothesis concerning responsibility to be evaluated merely on whether the evidence is maximally coherent with that hypothesis or its negation. In criminal trials, however, there is a presumption that convicting an innocent person is a harm to be avoided, even at the risk of letting a guilty person go. There are other ways in which reasonable doubt might be understood, for example, as requiring a higher probability, but problems abound with applying probability theory in legal

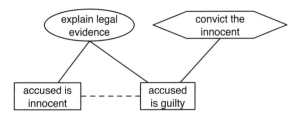

Figure 17.7
Map of reasonable doubt in criminal trials: acceptance of the hypothesis that the accused is guilty is inhibited by the negative value attached to convicting the innocent.

contexts (Thagard, 2004). As shown in figure 17.7, the hypotheses that the accused is innocent and that the accused is guilty are both evaluated with respect to how well they explain the evidence presented to the court, but the conclusion of guilt is somewhat inhibited by the connection to the negative value of convicting the innocent. Hence conviction requires stronger evidence than acquittal.

By analogy, we could understand scientific inference employing the harm principle of Rudner and Douglas as discouraging acceptance of a hypothesis if there is a risk of harm associated with it, as in the judgment that a chemical is not toxic. Figure 17.8 shows how this might work, introducing a presumption against accepting a hypothesis where the acceptance might lead to actions that cause harm. At this point, we have two thirds of the values shown in figure 17.5, as social values have once more been introduced into what figure 17.6 tries to mark as a purely epistemic exercise.

Disturbingly, however, the legal analogy seems to break down, because the British legal principles assume that convicting the innocent is a special harm, and do not worry explicitly about the risks associated with acquitting the guilty. In contrast, science seems to lack the asymmetry between accepting a hypothesis that causes harm and not accepting a hypothesis that might bring good. In some cases, rejecting a hypothesis that is potentially relevant to human needs can also cause harm, for example, if a drug is not approved that might have saved lives. Or consider climate change, as discussed in chapter 5. Accepting the hypothesis that human behavior is a major cause of climate change has potentially negative consequences if it leads to actions such as reducing climate emissions that limit economic growth and so increase poverty. However, the status quo

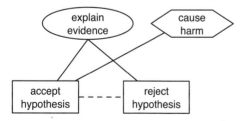

Figure 17.8
Map of caution in scientific inference: acceptance of the hypothesis is inhibited by the harm it might lead to.

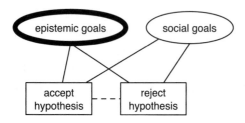

Figure 17.9
Social values affect the acceptance and rejection of hypotheses, but to a lesser extent than epistemic goals. The links from social goals to the nodes for accepting and rejecting the hypothesis may be positive or negative, depending on the risks and benefits of the hypothesis.

may be even more dangerous, if greater carbon emissions contribute to still more global warming, leading to disasters such as drought and flooding. So where do we end up? Perhaps figure 17.9 portrays the most appropriate values, allowing that social goals can play a small role, subordinate to epistemic goals, in determining what gets accepted and rejected.

What about personal values, whose incorporation into figure 17.9 would restore it to the same structure as figure 17.5? I can think of no good arguments why personal goals should play a role in hypothesis evaluation, which does not need the motivational and diversity concerns that support the inclusion of personal goals in the contexts of pursuit and application. The objective social aims of science are legitimate values relevant to establishing thresholds for acceptance and rejection of hypotheses, but personal values have no such legitimacy and can produce the biases discussed in the last section.

It is dangerous, however, to allow even a small role for social goals as in figure 17.9, because it opens the context of evaluation to serious biases. Examination of the ways in which social and personal goals have historically led to severe corruptions of scientific values may shed light on whether the value map in figure 17.9 is indeed the most appropriate one for the context of evaluation.

Biases in Scientific Inference

A defender of the ideal that evaluation of scientific hypotheses should concern only epistemic values can point to many disastrous incursions of

Table 17.1
Examples of interactions of good and bad social and epistemic values.

	Good epistemic values	Bad epistemic values
Good social values	1. Much of science, e.g., medical theories of infection	2. Lysenko's rejection of genetics
Bad social values	3. Nazi rocket science	4. Nazi rejection of Einstein

the social into the scientific, such as the Nazi rejection of Einstein's relativity theory as "Jewish science." Cognitive science can provide insights into the mental processes by which such biasing occurs, and thereby help to provide guidance about how to encourage people to follow normatively superior practices.

First we need a breakdown of ways in which epistemic and social values can interact. Table 17.1 displays four possibilities of ways in which good and bad values can interact. By "good epistemic values" I mean concepts such as truth and objectivity, shown in ovals in figure 17.2. By "good social values" I mean concepts such as needs and well-being, shown in ovals in figure 17.3. A great deal of science fits in the optimal quadrant 1, which combines good epistemic values with good social values. For example, much of modern medical science belongs here, as exemplified by the dramatic progress in understanding and treating infectious diseases: well-validated theories about how bacteria and viruses invade cells have contributed to a dramatic decrease in deaths from infections.

Quadrant 2 shows the rarer kind of case where good social values lead to the suspension of good epistemic values. The closest case I can think of is the Soviet Union's espousal of Lamarckian evolutionary theory under Lysenko, which was partly motivated by egalitarian concerns about social improvement: the hope was that intentionally altered environments could change the heritable nature of organisms ranging from plants to people. Unfortunately, this motivation led to neglect of the substantial amount of evidence supporting Mendelian genetics. The case of Lysenko was not, however, just about social goals, as it also involved the pursuit of personal goals such as power.

Other instances for quadrant 2 occur in some postmodernist critiques of science where justifiable concern with the mistreatment of people on the basis of their sex, race, and colonial status has led to the blanket rejection of Enlightenment values of truth and objectivity. Much alternative medicine also seems to fit in quadrant 2, where the desire to meet human needs for health leads people to ignore scientific procedures for evaluating medical treatments, such as controlled clinical trials.

Quadrant 2 encompasses the all-too-common cases where effective science is used for evil purposes. Nazi rocket science was scientifically strong, as shown by the recruitment of leading German scientists such as Werner von Braun by the Americans and Soviets after World War II. But the aim of the Nazi investment in rocketry was world domination, not satisfaction of universal human needs. Similarly, many other kinds of applied science from atomic weaponry to germ warfare employ epistemically good science in the service of aims inimical to general human welfare.

By far the most cognitively interesting and socially troubling kind of situation is shown in quadrant 4, where bad social values combine with bad epistemic values to produce atrocious results. Much pseudoscience fits in this category, where unscrupulous manipulators pursue their own personal goals of wealth and fame and promulgate ideas that are not only epistemically dubious but also harmful to those who adopt the ideas. A psychological example is the "secret" that people can get whatever they want just by wishing for it, an idea that has generated millions of book sales by beguiling people into overestimating the control they have over the universe. Climate change denial of the sort analyzed in chapter 5 also falls into this quadrant, when politicians in the pockets of oil companies distort science in order to pursue policies that in the long run hurt human welfare.

All of the cases in quadrants 2–4 arise from the cognitive-affective process of motivated inference that was discussed in chapter 5 concerning climate change. Recall that motivated inference is not just wishful thinking, but leads to more subtle distortions through selective use of evidence, not total abandonment. Social goals or personal goals eclipse epistemic goals without completely eradicating them, except in the most extreme cases where complete nihilism about evidence and reality is advocated. Figure 17.10 shows the general case where motivated inference allows social and personal values to eclipse epistemic ones, producing the pathologies shown in quadrants 2–4 of table 17.1.

Motivated inference is only one of the ways in which emotions can distort thinking; other affective afflictions include self-deception, weakness

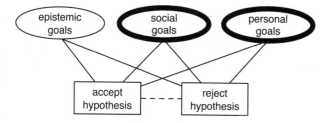

Figure 17.10
Motivated inference: eclipse of epistemic goals by social and/or personal goals. Thick ovals indicate higher emotional valences.

of will, conflicts of interest, depression, and manic exuberance (Thagard, 2007b, 2010a). There is an additional affective affliction of great importance that has received surprisingly little attention from psychologists and philosophers: countermotivated inference (Elster, 2007). Whereas in motivated inference people tend to believe what they desire, in countermotivated inference people tend to believe what they fear. This way of fixing beliefs sounds ridiculous, but it is very common in everyday situations such as jealous spouses, nervous parents, political paranoids, economic doomsayers, anxious academics, and hypochondriacs. Fear leads people to focus on scant evidence for a dangerous outcome, illegitimately increasing their belief in the outcome. I am now developing a neurocomputational model of how this kind of fear-driven inference works (Thagard, 2011).

Countermotivated inference can also occur in scientists and their critics. Science often leads to technologies whose consequences are hard to foresee, such as atomic power, recombinant DNA, and assisted reproduction. Sound science policy requires avoiding motivated inference in which the optimism of technologists sometimes leads them to ignore serious risks, but it also requires avoiding countermotivated inference in which people's ill-informed fears can lead them to be unduly convinced that a technology is dangerous. Examples might include excessive concerns about genetically engineered foods and about the Large Hadron Collider, which some commentators predicted would produce devastating black holes.

The susceptibility of human thinking to motivated and countermotivated inference naturally suggests swinging back to the traditional view that good reasoning should ignore emotion and instead employ normatively correct resources such as formal logic, probability theory, and utility theory. The problem with this reversion is that many important decisions in science and everyday life are hard to analyze using these tools, so that people naturally fall back on emotionally encoded values (Thagard, 2006a, ch. 2). There are no neat calculations that can tell us what to do about limiting climate change (ch. 5, this vol.) or teaching biology (ch. 14, this vol.). Hence the decisions that scientists and others need to make about what projects to pursue, what theories to accept, and what applications to enact will unavoidably have an emotional, value-laden aspect. Normatively, therefore, the best course is not to eliminate values and emotions, but to try to ensure that the best values are used in the most effective ways, avoiding affective afflictions.

One step toward effective reasoning is to make values explicit using cognitive-affective maps. Figure 17.10 shows generally how social and individual values can swamp epistemic ones, but the illumination of particular cases would require detailed identification of the specific values operating, including ones taken from figures 17.2–17.4. We could, for example, spell out what distorted social values led the Nazis to reject Einstein's theory. Cognitive-affective maps of values are only one tool useful for this task, supplementing the kind of propositional analysis of motivated inference that chapter 5 provided for climate change.

Another potential use for cognitive-affective maps is to display emotional conceptual changes like those described in Part IV. Someone who begins to appreciate the overwhelming evidence for Darwin's theory may change the valences associated with the concept of evolution and associated concepts. It would be interesting to map emotional conceptual changes that sometimes occur in scientists as well as students. These may concern concepts specific to particular theories, but also the most general epistemic and social values such as truth and well-being.

Conclusion

This chapter has used cognitive-affective maps to illuminate the legitimate and sometimes illegitimate roles that values play in scientific thinking. The maps are not just a diagramming technique, but also an exercise in computational cognitive science: they derive from the HOTCO artificial neural network model of how inferences are made, including ones that can be biased by emotional influences.

I have rejected the doctrine that science should be value free, but more importantly I have tried to show just how values do and should affect scientific decisions. Cognitive-affective maps have been useful for displaying the rich structure of epistemic, social, and personal values. Moreover, they show how all three kinds of values can legitimately contribute to decisions in the contexts of pursuit and application of scientific ideas. More subtly, the maps provide a way of seeing how social values can make a modest but valuable contribution in the context of evaluation, with inference taking into account the potential costs and benefits of accepting and rejecting hypotheses.

Just as important, cognitive-affective maps display how motivated inference can lead to distortions in scientific judgments when social or personal values play too large a role in the acceptance or rejection of scientific theories. Both good and bad social values can distort scientific inferences, in ways that cognitive science can explain by building models of mental mechanisms that simulate cases where minds work well, and also cases where minds fall short of normative standards because of affective afflictions.

Values are irremovable from science because they are irremovable from the minds of scientists. Attempts to eliminate values from science are not only futile but counterproductive, because the contexts of pursuit and application need values to guide decisions, and even the context of evaluation has a modest role for values. Understanding how minds work in all these contexts is crucial for developing both descriptive and normative accounts of inference. An epistemologist or educator who ignores cognitive science is like an engineer who ignores physics, or a medical doctor who ignores biology. Similarly, administrators, politicians, and others concerned with science policy need to appreciate what the cognitive science of science can tells us about the mental nature of values in science and their effects on many kinds of decisions.

What Are Concepts?

The discussion of conceptual change in Part IV assumed that concepts are an important part of scientific knowledge, but largely ignored the crucial question of what concepts are. Previous chapters mentioned many important science concepts such as *life, mind,* and *disease.* I have taken for granted the standard cognitive science assumption that concepts are mental representations, but have not provided a theory of what concepts are. This chapter draws on ideas of Chris Eliasmith (forthcoming) to argue that concepts, including scientific ones, are semantic pointers—neural processes with powerful semantic, syntactic, and pragmatic capabilities.

Most generally, concepts are representations corresponding to individual words that stand for classes of things; for example, the concept *car* corresponds to the word "car" that stands for the class of cars. The history of philosophy and cognitive science has witnessed many different interpretations of concepts, including:

• Concepts are abstract entities, for example, the forms of Plato.

• Concepts are copies of sense impressions, for example, the ideas of Hume (1888).

• Concepts are data structures that depict prototypes, for example, the frames of Minsky (1975).

• Concepts are distributed representations in neural networks, for example, the schemas of Rumelhart and McClelland (1986).

Another possibility is that concepts do not exist and should be eliminated from scientific discussions (Machery, 2009).

Rather than eliminating concepts, however, I think cognitive science needs to develop a robust theory of them that can contribute to the aims of all the constituent disciplines, including:

• Psychology: explain behavioral experiments about the use of words and address questions about innate versus learned knowledge.

• Philosophy: provide answers to epistemological questions about the nature of knowledge and meaning, and answers to metaphysical questions about what exists.

• Linguistics: explain how people produce and understand language.

• Neuroscience: explain neurological observations such as the loss of verbal ability in kinds of agnosia.

• Anthropology: explain cultural differences in language and categorization.

• Artificial intelligence: generate new data structures and algorithms for knowledge representation and language understanding.

More specifically, the cognitive science of science needs a theory of concepts for characterizing the structure and growth of scientific knowledge.

This chapter proposes a new approach to scientific concepts using the remarkable new theory of semantic pointers developed by Chris Eliasmith (forthcoming). A semantic pointer is a kind of neural representation whose nature and function is highly compatible with what is currently known about how brains process information. Semantic pointers are neural processes that (1) provide shallow semantics through relations to the world and other representations; (2) can be expanded to provide deeper semantics with relations to perceptual, motor, and emotional information; and (3) support complex syntactic operations. Semantic pointers are naturally represented mathematically by vectors in high-dimensional spaces, just as forces are physical processes that are naturally represented mathematically in two-dimensional spaces where the dimensions indicate magnitude and direction.

To take a simple example, consider the concept *car*. The concept *car* has important semantic functions such as referring to cars in the world, important syntactic functions such as the generation of sentences like "Electric cars help the environment," and important pragmatic functions such as helping to capture the intentions of speakers in particular contexts. Whereas traditional approaches to formal logic and linguistics take syntax as central and treat semantics and pragmatics as add-ons, Eliasmith's semantic pointer hypothesis shows how concepts construed as neural processes can simultaneously have syntactic, semantic, and pragmatic functions.

Neurons have synaptic connections that enable them to fire in patterns in response to inputs from other neurons. These patterns of firing can function syntactically as a result of the binding operations described in chapter 8: two concepts can be bound into a combined concept by convolution, and the same process can produce syntactic entities of great complexity (Eliasmith forthcoming). Such neural representations carry only partial, compressed meaning, but they can point to patterns of firing in neural populations that are much semantically richer by virtue of their relations to perceptual, motor, and emotional representations. The emotional associations of concepts contribute to the pragmatics of concept use. For example, if your past experiences with cars have led you to like them, then your future actions will be disposed to activities that use them. The concept *car* as a semantic pointer has an emotional component by virtue of connections between neural populations that carry the pointer and neural populations that encode goals. This component allows concepts to represent values as described in chapter 17. Hence concepts construed as semantic pointers participate in processes where brains are sensitive to contexts and goals, enabling concepts to contribute to the pragmatics of language and decision making.

In computer science, a pointer is a special kind of data type, with properties different from more familiar data types such as bits (0, 1), integers, and strings of characters. A pointer has a value that directs a program to a place in computer memory that stores another value, roughly the way that a street address directs a person to a place in a town where a house is located. Eliasmith's semantic pointers are a generalization of this idea, in that they can refer to multiple locations in a neural memory in a way that endows them with multifarious meanings. Semantic pointers have both a partial meaning that suffices for them to participate in syntactic operations such as forming sentences and entering into inferences, and also a deep meaning that can be accessed by reference to multiple memory locations with perceptual, motor, and emotional information.

Semantic pointers are analogous to compressed computer files that throw away much information but are still adequate for their intended purpose. For example, the song files that store music in applications such as iTunes and mp3 players drop much of the information contained in an analog recording, but still contain the digital data needed for a speaker to reproduce music satisfying to almost all listeners. Similarly, semantic pointers drop much sensory detail about the external world, but are still able to participate in the many kinds of inference involved in perception, motor control, and even high-level reasoning. Like compressed audio files, semantic pointers are more efficient to transport and manipulate than uncompressed information, which pointers can regenerate when it is needed for deeper processing.

Mathematically, semantic pointers can be described as vectors in a high-dimensional space. A vector is a mathematical structure that represents two or more quantities. A simple example is a vector that represents a car traveling at 100 kilometers per hour headed east, which is a two-dimensional vector that can be expressed as a structure (100, 90) where 100 is the speed in kilometers and 90 is the angle in degrees. The space in which semantic pointers operate includes hundreds of dimensions derived from information relevant to their verbal, perceptual, or motor uses. The mathematical representation of semantic pointers is very useful for exploring their syntactic and semantic functions; but such explorations are highly technical, so the rest of this chapter will stick to the mechanistic terminology of patterns of firing in neural populations. A more general defense of the desirability of thinking of concepts and other mental representations as neural processes can be found in Thagard (2010a).

For the semantic pointer interpretation of concepts to contribute to the cognitive science of science, it should help to provide answers to questions such as the following:

1. How are scientific concepts meaningful, especially theoretical ones that are distant from sense experience?

2. How can scientific concepts contribute to scientific explanations?

3. How can scientific concepts contribute to discoveries and conceptual change?

4. How can scientific concepts contribute to the practical goals of science?

I will now show the relevance of semantic pointers to these questions by considering the nature of three fundamental scientific concepts from physics, chemistry, and biology: *force, water,* and *cell.*

Force

The concepts *force*, *water*, and *cell* are central in their respective sciences, and all have multimodal aspects that fit well with the semantic pointer view of concepts. By "multimodal" I mean that these concepts are mental representations that are not only verbal but also contain information tied to sensory, motor, and emotional modalities.

Consider first the concept of force, which is fundamental in modern physics (Jammer, 1957). Physics students are taught about gravitational force and move on to learn about other forces such as electromagnetic, frictional, viscous, adhesive, chemical, molecular, and nuclear. According to current physical theory, there are four fundamental forces: gravitational, electromagnetic, and the strong and weak forces that operate at the atomic level. Commonly, a force is defined as an influence that causes a body to undergo a change in speed, direction, or shape, but this definition is not very helpful, since *influence* seems to be much the same concept as *force*, and the concept of *cause* is notoriously difficult to define. Introductory physics texts often define a force as a push or a pull, which seems anthropomorphic rather than scientific.

Positivists such as Ernst Mach worried that there was something illegitimate about the concept of force, despite its centrality in Newtonian physics. Following the empiricist philosophy of Locke, Berkeley, and Hume, they assumed that the meaning of concepts derives from sense experience, which made Newton's idea of force suspect, since force as characterized by Newton's laws is not observable. Force equals mass times acceleration, but unlike mass and speed, force is not directly measurable. Science educators remain puzzled about how to teach students about the nature of force (Coelho, 2010).

The mysteriousness of the meaning of *force* dissipates if one considers both the history of the concept and the semantic pointer interpretation of concepts. According to Jammer (1957, p. 7), the concept of force originated in familiarity with human will power and muscular effort and became projected onto inanimate objects as a power dwelling in them. Initially, this power was construed mentally, so that the force in objects was viewed as depending on spirits or divine action. By the seventeenth century, however, the mental construal of force had dropped out through the influence of Kepler, Galileo, and others.

In contrast to the association with mental activity, the muscular effort association of force survives in the common idea that a force is a push or a pull. According to Jammer (1957, p. 17):

The idea of force, in the prescientific stage, was formed most probably by the consciousness of our effort, spent in voluntary actions, as in the immediate experience of moving our limbs, or by the consciousness of the feeling of a resistance to be overcome in lifting a heavy object from the ground and carrying it from one place to another.

How do we know what pushes and pulls are? They are not easily definable in terms of other words, but rather involve visual, tactile, and kinesthetic sensations. When I push a box, I can see and feel myself in contact with the box, and, just as important, I have the kinesthetic sense of my arms and the rest of my body moving. Pushing is not captured by a single kind of sense experience, but rather by a combination of visual, tactile, and kinesthetic sensations. Pulling has a different combination involving another set of motions of muscles and body parts. People's concepts of pushes and pulls are not abstract or purely verbal entities, but rather amount to representations involving several sensory modalities. The fact that pushes and pulls require a combination of modalities shows that they are not simply copies of sense impressions as Hume assumed.

On the semantic pointer interpretation of concepts, mental representation of pushes and pulls is a neural process involving firing in populations of spiking neurons linked to brain areas capable of visual, tactile, and kinesthetic representations. The relevant areas are mainly in the visual cortex, somatosensory cortex, and motor cortex, respectively. The highly distributed neural population that represents pushes combines these modalities together using the sort of binding discussed in chapter 8. The deep semantics of the concept of pushing comes from the multiple sensory modalities that perceive pushing by oneself or by others, although particular uses of the concept of pushing need not access the full range of sensory activations. Hence, the neural population most active when we think of pushing can provide a compressed representation whose shallow semantics involves correlations with other neural populations for kinds of objects related to pushing, such as doors and people. Kinesthetic representations of force can also arise from experiences with magnets, including the feeling of pulling that comes when a magnet attracts metal, but also pushing when two magnets repel each other.

Obviously, concepts in modern physics go far beyond muscular representations of pushes and pulls. Newton's *Principia* contains the beginning of the modern concept of force, extending ideas developed by Kepler and Galileo. Here are Newton's famous three laws of motion (Jammer, 1957, p. 123):

I. Every body continues in its state of rest, or of uniform motion in a right line, unless it is compelled to change that state by force impressed upon it.

II. The change of motion is proportional to the motive force impressed; and is made in the direction of the right line in which that force is impressed.

III. To every action there is always opposed an equal reaction; or the mutual actions of two bodies upon each other are always equal, and directed to contrary parts.

Here force is not restricted to the actions of human bodies, but can be attributed to other actions such as the motion of planets. Newton's concept of force, however, did not come out of the blue, but built on his sensory understanding of force derived from experiences with pushes and pulls.

Newton expressed his second law in words, but it naturally has a mathematical expression: $F = ma$ or $F = d(mv) / dt$. The reason that mathematical concepts are so useful is that they provide a compression similar to what happens with semantic pointers. It is hard to think abstractly about what forces are as causes of acceleration and as pushes or pulls, but mathematical representations such as vectors discard the many associations of deep semantics and provide symbolic representations that can be manipulated syntactically in mathematical operations such as proofs and calculations. Numbers and variables have the same effect: it is mentally easier to manipulate "7" than "7 things," or "x" rather than "an unknown number." Mathematical representations dramatically reduce the load on short-term working memory, just as semantic pointers do. Hence it is useful to represent forces by vectors in two-dimensional spaces, and semantic pointers as vectors in higher-dimensional spaces.

Thus the concept of force has its origins in multimodal sensory experience, but through verbal and mathematical theories can be expanded into the amazingly rich kind of representation found in the minds of physicists.

But the accretion of large amounts of verbal and mathematical information to the concept of force does not overturn its connections with sensorimotor experience.

Water

For a chemical example, consider the concept *water*. Like *force, water* has a substantial prescientific meaning, and the semantic pointers that operate in ordinary people's brains tie water to many sensory experiences, including taste and smell as well as vision and touch. Sound is also tied to the concept of water through the familiar rhythms of waves, and there is also kinesthetic experience of water familiar through wading and swimming. Thus the neural processes that encode the concept *water* point to experiences in many modalities.

Like force, however, water can take on diverse theoretical roles, extensively reviewed by Grisdale (2010). The early Greek philosopher Thales proposed that water is the basic element out of which everything else is formed. Aristotle disagreed and proposed that there were four other elements besides water: earth, air, fire, and aether. Water was viewed as an element for more than two thousand years until the 1700s, when researchers such as Lavoisier provided evidence that water is actually a compound rather than an element, consisting of hydrogen and oxygen. The association of water with H_2O is now part of the conceptual structure of anyone with rudimentary instruction in science. This part of the representation of water is largely verbal, although images of water showing two hydrogen atoms joined to an oxygen atom are also available. Today, quantum theory explains how bonds form between hydrogen and oxygen atoms through electron attractions.

Despite all this theoretical progress, the concept of water for scientists retains the multimodal associations it has for ordinary people. The concept of water has changed enormously from the ancient mythological views that tied different forms of water to various gods. The view of water as an element has been replaced by a far more explanatorily successful view of water as a specific kind of compound. Nevertheless, it is legitimate to say that the concept of water was changed rather than abandoned because of the retention of multimodal information carried by the semantic pointers of people from the ancient Greeks to today.

Grisdale's (2010) discussion of modern conceptions of water refutes a highly influential thought experiment that the meaning of water is largely a matter of reference to the world rather than mental representation. Putnam (1975) invited people to consider a planet, Twin Earth, that is a near duplicate of our own. The only difference is that on Twin Earth water is a more complicated substance XYZ rather than H_2O. Water on Twin Earth is imagined to be indistinguishable from H_2O, so people have the same mental representation of it. Nevertheless, according to Putnam, the meaning of the concept *water* on Twin Earth is different because it refers to XYZ rather than H_2O. Putnam's famous conclusion is that "meaning just ain't in the head."

The apparent conceivability of Twin Earth as identical to Earth except for the different constitution of water depends on ignorance of chemistry. As Grisdale (2010) documents, even a slight change in the chemical constitution of water produces dramatic changes in its effects. If normal hydrogen is replaced by different isotopes, deuterium or tritium, the water molecule markedly changes its chemical properties. Life would be impossible if H_2O were replaced by heavy water, D_2O or T_2O; and compounds made of elements different from hydrogen and oxygen would be even more different in their properties. Hence Putnam's thought experiment is scientifically incoherent: If water were not H_2O, Twin Earth would not be at all like Earth.

This incoherence should serve as a warning to philosophers who try to base theories on thought experiments, a practice I have criticized in relation to concepts of mind (Thagard, 2010a, ch. 2). Some philosophers have thought that the nonmaterial nature of consciousness is shown by their ability to imagine beings (zombies) who are physically just like people but who lack consciousness. It is entirely likely, however, that once the brain mechanisms that produce consciousness are better understood, it will become clear that zombies are as fanciful as Putnam's XYZ. Just as imagining that water is XYZ is a sign only of ignorance of chemistry, imagining that consciousness is nonbiological may well turn out to reveal ignorance rather than some profound conceptual truth about the nature of mind. Of course, the hypothesis that consciousness is a brain process is not part of most people's everyday concept of consciousness, but psychological concepts can progress just like ones in physics and chemistry (Thagard, forthcoming-b).

Zombies aside, we should draw the lesson that a theory of scientific concepts should be based on scientific evidence, not thought experiments. The semantic pointer interpretation of concepts shows how to integrate the prescientific, multimodal understanding of concepts based on vision and other sensory modalities with the theoretical developments about the constitution of elements provided by chemistry and quantum theory. Hence concepts can undergo dramatic change while retaining sensory continuity.

Cell

Biology also undergoes conceptual change, and no concept in biology is more central than *cell*, representing the smallest unit of life. This concept is multimodal in ways that make sense from the perspective of the semantic pointer view of concepts. Unlike force and water, people have no direct sensory experience of cells, which were first observed around the 1660s when Robert Hooke (1665) used a new instrument, the microscope, to examine a piece of cork. He identified small bounded areas in cork that he variously called pores, boxes, and cells, adapting the latter term from the Latin word for a small room. It was only in the 1830s that theories of the formation and function of cells were developed, thanks to improved observations using microscopes with achromatic lenses (Bechtel, 2006). Detailed observations of the structure of the cells only became possible in the 1940s through the development of the electron microscope that could identify small structures such as mitochondria.

In all three main stages of development, the concept of a cell had a substantial visual component, even though plant cells (unlike rooms) are not directly observable. Hooke's microscope enabled him to see similarities between cells and objects (pores, boxes, room) that are observable and for which people already had concepts. The new concept of cell arose by a process of visual analogy, combining the observation of new structures within plants with past observations of familiar objects. Hence Hooke's concept of a cell as a component of organic material such as corks and carrots was in part a pointer to the visual representation that he acquired through his own observations and that others could acquire through looking at the elegant illustrations in his book *Micrographia*. Much richer visual representations arose with better microscopes leading to the beautiful color illustrations of cells in modern textbooks and videos of mitosis that can easily be found on the Web. Neurons are nerve cells that became observable with optical microscopes in the 1870s thanks to improved staining techniques; the fine structure of neurons as synaptic connections became observable in the 1940s via electron microscopes. In this way, the visual aspect of cells carried in the deep semantics of the semantic pointer for *cell* allowed the concept to retain some continuity despite major theoretical changes.

I have argued that three central scientific concepts—*force*, *water*, and *cell*—can be understood as semantic pointers. All three have a substantial multimodal aspect accommodated by the expansion capabilities of semantic pointers, while retaining the capacity to serve as symbols in theories in physics, chemistry, and biology. After further discussion of meaning, I will argue that the semantic pointer hypothesis applies to concepts generally.

Meaning

Philosophers have long debated about the source of the meaning of concepts. Here are five possible answers to the question of why concepts are meaningful:

1. Concepts are meaningful because of sense experience (Hume, 1888).

2. Concepts are meaningful because they are innate.

3. Concepts are meaningful because they refer to things in the world (Putnam, 1975).

4. Concepts are meaningful because they have functional, procedural roles in relation to other concepts (Harman, 1987; Miller & Johnson-Laird, 1976).

5. Concepts are meaningful because they are used socially in communication (Wittgenstein, 1968).

The semantic pointer view of concepts does not force a choice among these answers construed as alternative theories, but rather shows how multiple processes can contribute to the meaning of concepts. The five sources of meaning are not completely distinct, because sensory experience, innateness, reference, and social use can all contribute to functional roles, and reference through interaction with the world can contribute to sense experience. Nevertheless, it would be futile to try to reduce the five sources of meaning to a smaller, fundamental set, ignoring the other kinds of interactions that provide the full relational range of meaning.

I will call this comprehensive view the *multirelational* theory of concept meaning because it proposes that the meaning of a concept derives from many processes that can affect patterns of neural firing that constitute semantic pointers. I avoid talking of the *content* of concepts because that term misleadingly suggests that meaning is a thing rather than a relational process. Similarly, we should not talk of the meaning of a sentence as a content or a proposition, which philosophers often construe as an abstract entity for whose existence there is no evidence. The conceptual change of thinking of meaning as a process rather than a thing is analogous to the important historical shift of thinking of mass as a quantity like weight to thinking of it as the result of a process relating multiple objects.

To connect the multirelational view of meaning with the theory of concepts as semantic pointers, I need to show how semantic pointers relate to sense experience, innateness, reference, functional role, and social uses. The neural processes that constitute concepts are often causally correlated with sense experience, most obviously in sensory concepts such as *heavy* and *blue*. Neural populations can encode sensory phenomena because particular neurons become tuned to different aspects of sensory inputs. However, the semantic pointer view of concepts is not restricted to a narrowly empiricist view because the synaptic connections that generate patterns of firing may be the result of internal interactions with other neural populations, not directly with sensory inputs.

Moreover, some of the synaptic connections that generate patterns of firing may be innate, as is perhaps the case with a few core concepts like *object* or *face*. The issue of the extent of innateness of concepts is highly controversial, with current views ranging from ones that minimize innateness in favor of powerful learning capabilities (e.g., Quartz & Sejnowski, 1997, 2002; Elman et al., 1996) to ones more inclined to posit innate concepts (e.g., Carey, 2009). The semantic pointer hypothesis is neutral on the question of innateness, and is compatible both with meaning arising from sense experience and with meaning arising from innate connections instilled by natural selection.

Perhaps the concept *face* is innate in humans, as suggested by the finding that infants orient to and respond to faces shortly after birth (Slater & Quinn, 2001). This concept is clearly not just verbal, as the language resources of newborns are at best limited, but involves visual representations that indicate the appropriate structure of eyes, nose, and mouth. Face representation is dynamic, as infants respond to changes such as smiles. Supposing that the concept of face is innate requires infants to be born with neural populations that detect features of faces and the overall configuration that signifies a face. But the concept, construed as a semantic pointer, can operate also at a symbolic level that can generate rulelike behaviors such as: If someone makes a face, then make a face back. To produce these results, all that is needed is to have brain development put in place before birth a set of neurons and synapses that will generate the appropriate firing patterns when the infant receives sensory stimulation. Because such structures are no different from the ones that the semantic interpretation of concepts assumes can be learned from experience, this interpretation can accommodate meaning arising from both sense experience and innateness.

The semantic pointer interpretation makes sense of the origins of sensory concepts, but it can also recognize that for many concepts the correlations with sense experience are highly remote. The remoteness is particularly acute with theoretical concepts in science such as *electron*, *virus*, and *black hole*. Positivist philosophers worried about how such concepts could be meaningful when they are not translatable into sense experience, but they never justified their contention that all meaning must derive just from the senses. Concepts can also gain their meaning from interactions with other concepts, which need not be definitional: very few terms outside mathematics have strict definitions, as the above examples of force, water, and cell confirm.

It is easy to see how semantic pointers can get some of their meaning from other semantic pointers, both through low-level processes like spreading activation and stronger more rulelike associations. For example, the meaning of the concept *car* derives in part from its association with kinds like *vehicle* and with parts like *engine*. However, whereas the semantic pointer for *car* can be expanded into pictures, sounds, and smells of cars, there is no similar sensory expansion of *electron* and *black hole*, because we have no senses or instruments for observing them. Nevertheless, such concepts can be meaningful because the neural processes that constitute them have systematic causal relations to other concepts, including ones tied to sensory experience. For example, we have reason to believe that electrons exist because of their causal effects on many observable results, such as lamps going on. The concept *electron* is then associated neurally with the observable concept *lamp*, allowing semantic pointers to get part of their meaning from relations to other concepts.

Whereas empiricist philosophers maintained that concepts get meaning from sense experience, realists see meaning as emanating from relations to the world. From the perspective of a neuropsychological view of how perception works, the sensory view of meaning fits well with the referential view of meaning, because interactions with the world are via the senses. My concept *car* gets much of its meaning from my sense experiences of cars, but the physiology of perception tells us that this sense experience is often the result of causal interactions with cars. For example, light reflects off a car into my eyes and produces my visual experience of a car. The motions of a car engine generate sounds waves that stimulate my ears to produce auditory experiences of a car. Hence reference can be seen to be part of the meaning of *car*, construed as a semantic pointer, via the causal interactions between people's brains and objects in the world. People need not be passive recipients in such interactions, but can use their bodies to manipulate the world in ways that generate new sensory experiences, as when I turn a key to start a car and produce new sounds and sights. For more on neurosemantics, see Eliasmith (2005b) and Parisien and Thagard (2008).

There is one more source of the meaning of concepts that must be mentioned. The activation of concepts is not just the result of interactions with the world, but for language-using beings is often also the result of interactions with other people. The associations between semantic pointers described in relation to functional interactions often arise because of communication that people have with others. I need never have seen a penguin to have a concept of penguin, not just because I have seen pictures of penguins, but also because I have heard other people talk of penguins. Conversations generate neural activity, sometimes leading to the acquisition of new concepts, as in the rapid vocabulary increase found in children and other learners. The semantic pointers of one person make possible uses of language that are perceived by other people, and thereby can lead to altered neural processes in the listener, including new concepts.

Thus the semantic pointer interpretation of concepts is consistent with the multirelational view of meaning as potentially deriving from *all* of sense experience, innateness, functional roles, reference, and social use. It should be clear that this view has a ready answer to various conundrums that have plagued the view that meaning belongs to abstract symbols. The symbol-grounding problem that arises with purely computational systems with no relation to the world is clearly not a problem for semantic pointers, which have causal links with sense experience and reference. Such links need not be special to brains, as robots can have them too (Parisien & Thagard, 2008). Semantic pointers are also consistent with the moderate embodiment thesis discussed in chapter 4, because their multimodal expansions operate via the full range of bodily senses, including kinesthesia that affects concepts like *force* and *water*.

However, the semantic pointer view is incompatible with the extreme embodiment thesis that throws out ideas about representation altogether. I see as one of the major virtues of the semantic pointer hypothesis that it shows how cognition can be grounded, embodied, *and* representational. Looking back to chapter 4, it should be clear how semantic pointers can support mental models that have syntactic as well as semantic properties.

What Concepts Are

The sections on force, water, and cells showed that three central scientific concepts can naturally be understood as semantic pointers, but it would be hasty to generalize from a few examples to all scientific concepts, let alone all concepts in other domains. The best way to establish generally that concepts are semantic pointers, that is, neural structures and processes with the syntactic, semantic, and pragmatic properties described above, would be to show that this supposition explains many results of psychological experiments. There are thousands of experiments about concepts, and ideally semantic pointers could be used to simulate the full range of results. In place of that overwhelming task, it would be desirable to show that semantic pointers can display the kinds of behaviors that have been used by proponents of various psychological theories to explain experimental findings.

Currently in cognitive science, the three main competing interpretations of the nature of concepts construe them as prototypes, sets of exemplars, or parts of explanatory theories (Machery, 2009; Murphy, 2002;

Thagard, 2005a). A unified theory of concepts would have to show that the phenomena supporting all three interpretations can be explained by the same mechanism, in this case semantic pointers. My colleagues and I plan to use computer simulations of neural networks to show that this unification is feasible, but in advance of such results I can speculate about how semantic pointers might be relevant to understanding the full range of properties of concepts described by available theories.

Since the 1970s, many philosophers, psychologists, and computer scientists have advocated a view of concepts as "prototypes," which are mental representations that specify typical rather than defining properties. Prototypes are more flexible than definitions and there are experimental reasons to think that they give a better account of the psychology of concepts. However, they may not be flexible enough, so some psychologists have claimed that people do not actually store concepts as prototypes, but rather as sets of examples. This claim is called the "exemplar" theory of concepts. The third major account of concepts currently discussed by psychologists is called the "knowledge" view or sometimes the theory-theory. This view points to the large role that concepts play in providing explanations.

Depending on which theory of concepts one adopts, the concept of force, for example, would be variously construed as a representation of typical properties of forces, as a set of examples of forces, or as a theoretical structure providing explanations in terms of forces. Thagard (2010a) contends that these are not competing theories of concepts but merely point to different aspects of them that can be unified using sufficiently rich neurocomputational ideas.

Semantic pointers seem to have the desired richness. A neural network such as those trained by back propagation can acquire its connection weights by exposure to a large number of examples and thereby have the ability to access an approximate representation of them. The semantic pointer, however, need not have the full representation of those examples, which would make it unwieldy and incapable of managing the prototype and explanatory functions of concepts. But the semantic pointer can point (through neural connections) to locations in the brain where information about multiple examples is stored multimodally.

For more efficient processing, a semantic pointer can encode a set of typical features that apply to a class of things, in just the way that typical trained neural networks can encode prototypes. A neural population that fires in a recognizable fashion when presented with particular sets of features need not treat those features as necessary or sufficient conditions for the application of a concept. Rather, the neural network responds probabilistically to the presented features and makes an approximate guess about which concept an object possessing a set of features falls under.

Finally, whereas exemplar and prototype theories of concepts have difficulty accounting for their explanatory uses, it is easy to see how semantic pointers can figure in explanations by virtue of their symbol-like ability to figure in approximate causal rules. In accord with the view of causality advocated in chapter 3, the basic human schema for causality is a sensory-motor-sensory pattern that is either innate or acquired within a few months of birth. Through learning, this schema gets generalized into an observation-manipulation-result pattern that goes beyond direct sensory experience and can be further expanded into the technical realm that comes with such mathematical advances as probability theory and Bayesian networks. People acquire rules of the form concept-cause-result, which could be expressed as a rulelike structure of the form "If x falls under a concept C then it has behavior B." For example, if we want to explain why an animal quacked, we could use the causal knowledge that ducks produce quacks.

I am assuming that all of the theoretical uses of concepts—as sets of exemplars, as prototypes, and as explanations—retain access to the multimodal, perceptual representations of concepts. That retention does not fall back into the empiricist view of concepts as simply derived from sense experience, which would not handle the prototype and explanatory uses of concepts, but it shows that even for those uses semantic pointers can retain contact with the empirical.

This section has been highly speculative and should be taken as tentative until simulations can be produced that show that semantic pointers as implemented in Eliasmith's Neural Engineering Framework can account for all the major characteristics of concepts as revealed in psychological experiments. As chapter 1 described, the purposes of computer simulations include showing that a proposed mechanism is coherent enough to be implemented in a program and powerful enough to generate the desired behaviors. The proof is in the programming.

An alternative to a unified theory of concepts is the skeptical conclusion of Machery (2009) that exemplars, prototypes, and explanations employ very different kind of representations, so cognitive science should simply eliminate concepts from theoretical discussions. Such elimination is unnecessary, however, if semantic pointers can show how concepts, construed as neural processes, can function under different circumstances in ways prescribed by all three ways of understanding them.

According to Machery (2009), philosophers look to concepts to serve a very different purpose from the aim of psychologists to explain experimental results about how people classify things. He says that the purpose of concepts in most contemporary philosophy is to explain how people can have "propositional attitudes" such as beliefs, desires, and hopes. Such mental states are taken as relations between a person and an abstract entity called a proposition. I have argued elsewhere that the philosophical idea of propositional attitudes is a mistake and should be supplanted with ideas about mental representation taken from cognitive science (Thagard, 2008a). What remains of the abstract philosophical project of understanding propositional attitudes is the question of how concepts, construed naturalistically, can figure in more complex kinds of representations analogous to sentences that make claims about the world.

For this purpose, the semantic pointer interpretation of concepts is highly useful, as there is now a developed theory of how concepts can combine computationally into more complex structures of great complexity. Even embedded sentences such as "That Mary insulted John caused Mary to dislike John" can be encoded by vectors and implemented in neural networks (Eliasmith, forthcoming; Eliasmith & Thagard, 2001; Plate, 2003). We have already seen that semantic pointers can support the many sources of meaning that philosophers have looked for at the concept level, and they can be combined in ways that support the multirelational nature of meaning.

Conclusion

In the late 1970s, I wrote a paper called "Scientific Theories as Frame Systems," which I never published, even though it is one of the best I ever wrote. The reason I never submitted it for publication is that my ideas about applying the frame theory of Minsky (1975) to understanding scientific knowledge began to seem to me rather vague once I learned computer programming. Many of the ideas from that paper appear in more specific form in my 1988 book *Computational Philosophy of Science*. This chapter has had the same aim as that book and the unpublished paper on frames—to use the resources of cognitive science to go beyond usual approaches to understanding the growth and structure of scientific knowledge. I think that the semantic pointer view of concepts is more powerful than Minsky's frame account, in that it elegantly and precisely accommodates multimodal, exemplar, and explanation aspects of concepts, in addition to prototype aspects.

A major advantage of the semantic pointer view of scientific concepts is that the sensory aspects of force, water, and cells can be retained in their deep, expanded semantics without encumbering their many theoretical functions. Another advantage is the clarification of how concepts like *force*, *water*, and *cell* can change dramatically with new scientific theories while remaining identifiably continuous with earlier concepts. Thomas Kuhn's (1962) theory of scientific revolutions seemed to have the radical implication that meaning change across paradigm shifts is so dramatic that rational evaluation is impossible. Semantic pointers show how the development of concepts can display continuities in the face of dramatic theoretical change, so that scientific developments can be rational even when revolutionary (Thagard, 1992). This combination of change and continuity requires a complex theory of meaning that takes into account sensory experience, innateness, reference, functional roles, and concept uses.

Although concepts like *force, water,* and *cell* retain some continuity based on multimodal semantics, it must also be recognized that sometimes theoretical advances do require rejection of assumptions based on sensory experience. We now recognize that, contrary to the original conception, force does not require will and can operate at a distance. Water is not just the familiar liquid and solid, but can also occur in gaseous form and be decomposed into oxygen and hydrogen. Whereas Hooke saw cells as inert walls, we now conceive of them as living entities that can reproduce. It is striking that the five elements of Aristotle have all been dramatically reclassified through theoretical advances. Water, earth, and air have all been reclassified as compounds rather than elements; fire is neither an element nor a compound, but a process resulting from rapid oxidation; aether does not exist. A theory of concepts needs to be rich enough to allow for both conceptual continuity and dramatic kinds of conceptual change: the semantic pointer interpretation is up to the challenge.

Another advantage of the semantic pointer view of concepts is that it is part of a general cognitive architecture that surmounts many of the divisions that have arisen in cognitive science in the past few decades. Contrary to criticisms of the mainstream computational-representational approach in cognitive science, Eliasmith's (forthcoming) semantic pointer architecture shows how thinking can be all of the following: psychological and neural, computational and dynamical, symbolic and subsymbolic, rule-governed and statistical, abstract and grounded, disembodied and embodied, reflective and enactive (action-oriented). Achieving these reconciliations, however, requires expansion of the theoretical resources of traditional cognitive science to include the full range of syntactic, semantic, and pragmatic capabilities of semantic pointers.

The arguments in this chapter have not definitively established the semantic pointer interpretation as *the* correct view of scientific concepts or concepts in general. But they show the general power of the neural view of concepts that chapter 8 used to explain creative conceptual combination. I have described how important science concepts like *force*, *water*, and *cell* can be understood as semantic pointers. More generally, I have suggested how the exemplar, prototype, and explanatory aspects of concepts can all be understood in terms of semantic pointers. Hence the semantic pointer interpretation of concepts makes a potentially large contribution to the cognitive science of science.

References

ACM. (2002). Crossroads: The ACM Magazine for Students. Retrieved February 15, 2011, from http://xrds.acm.org.

Adler, R. E. (2004). *Medical firsts: From Hippocrates to the human genome.* Hoboken, NJ: John Wiley & Sons.

Alexander, R. C., & Smith, D. K. (1988). *Fumbling the future: How Xerox invented, then ignored, the first personal computer.* New York: Morrow.

Allchin, D. (1996). Points east and west: Acupuncture and comparative philosophy of science. *Philosophy of Science, 63*(supplement), S107–S115.

Amir, R. E., Van den Veyver, I. B., Wan, M., Tran, C. Q., Francke, U., & Zoghbi, H. Y. (1999). Rett syndrome is caused by mutations in X-linked MECP2, encoding methyl-CpG-binding protein 2. *Nature Genetics, 23*(2), 185–188.

Andersen, H., Barker, P., & Cheng, X. (2006). *The cognitive structure of scientific revolutions.* Cambridge: Cambridge University Press.

Anderson, J. R. (1983). *The architecture of cognition.* Cambridge, MA: Harvard University Press.

Anderson, J. R. (1993). *Rules of the mind.* Hillsdale, NJ: Erlbaum.

Anderson, J. R. (2007). *How can the mind occur in the physical universe?* Oxford: Oxford University Press.

Anderson, J. R. (2010). *Cognitive psychology and its implications* (7th ed.). New York: Worth.

Anderson, J. R., Bothell, D., Byrne, M. D., Douglas, S., Lebiere, C., & Qin, U. (2004). An integrated theory of the mind. *Psychological Review, 111,* 1030–1060.

Anderson, R. D. (2007). Teaching the theory of evolution in social, intellectual, and pedagogical context. *Science education, 91,* 664–677.

Arrhenius, S. (1896). On the influence of carbonic acid in the air upon the temperature of the ground. *Philosophical Magazine and Journal of Science, 41*(April), 237–276.

Arthur, W. B. (2009). *The nature of technology: What it is and how it evolves*. New York: Free Press.

Asheim, B. T., & Gertler, M. S. (2005). The geography of innovation: Regional innovation systems. In J. Fagerberg, D. C. Mowery, & R. R. Nelson (Eds.), *The Oxford handbook of innovation* (pp. 291–317). Oxford: Oxford University Press.

Atran, S., & Medin, D. L. (2008). *The native mind and the cultural construction of nature*. Cambridge, MA: Cambridge University Press.

Au, T. K. (1983). Chinese and English counterfactuals: The Sapir-Whorf hypothesis revisited. *Cognition, 15*, 155–187.

Bacon, F. (1960). *The New Organon and related writings*. Indianapolis: Bobbs-Merrill.

Baillargeon, R., Kotovsky, L., & Needham, A. (1995). The acquisition of physical knowledge in infancy. In D. Sperber, D. Premack, & A. J. Premack (Eds.), *Causal cognition: A multidisciplinary debate* (pp. 79–116). Oxford: Clarendon Press.

Barnes, B., Bloor, D., & Henry, J. (1996). *Scientific knowledge: A sociological analysis*. Chicago: University of Chicago Press.

Barsalou, L. W., Simmons, W. K., Barbey, A. K., & Wilson, C. D. (2003). Grounding conceptual knowledge in modality-specific systems. *Trends in Cognitive Sciences, 7*, 84–91.

Bechtel, W. (2006). *Discovering cell mechanisms: The creation of modern cell biology*. New York: Cambridge University Press.

Bechtel, W. (2008). *Mental mechanisms: Philosophical perspectives on cognitive neuroscience*. New York: Routledge.

Bechtel, W., & Abrahamsen, A. A. (2005). Explanation: A mechanistic alternative. *Studies in History and Philosophy of Biology and Biomedical Sciences, 36*, 421–441.

Bechtel, W., & Richardson, R. C. (1993). *Discovering complexity*. Princeton: Princeton University Press.

Bechtel, W., & Richardson, R. C. (1998). Vitalism. In E. Craig (Ed.), *Routledge Encyclopedia of Philosophy* (pp. 639–643). London: Routledge.

Bentall, R. (2004). *Madness explained: Psychosis and human nature*. London: Penguin.

Bermudez, J. L. (2010). *Cognitive science: An introduction to the science of the mind*. Cambridge: Cambridge University Press.

Berton, O., McClung, C. A., Dileone, R. J., Krishnan, V., Renthal, W., Russo, S. J., et al. (2006). Essential role of BDNF in the mesolimbic dopamine pathway in social defeat stress. *Science, 311*(5762), 864–868.

Blackwell, W. H., Powell, M. J., & Dukes, G. H. (2003). The problem of student acceptance of evolution. *Journal of Biological Education, 37*, 58–67.

Blanshard, B. (1939). *The nature of thought* (Vol. 2). London: George Allen & Unwin.

Blasdel, G. G., & Salama, G. (1986). Voltage sensitive dyes reveal a modular organization in monkey striate cortex. *Nature, 321*, 579–585.

Bloom, A. (1981). *The linguistic shaping of thought: A study of the impact of language on thinking in China and the West.* Hillsdale, NJ: Erlbaum.

Bobrow, D. G., & Collins, A. (Eds.). (1975). *Representation and understanding: Studies in cognitive science.* New York: Academic Press.

Boden, M. (2004). *The creative mind: Myths and mechanisms* (2nd ed.). London: Routledge.

BonJour. L. (1985). *The structure of empirical knowledge.* Cambridge, MA: Harvard University Press.

Bowden, E. M., & Jung-Beeman, M. (2003). Aha! Insight experience correlates with solution activation in the right hemisphere. *Psychonomic Bulletin & Review, 10,* 730–737.

Breakenridge, R. (2008). What is it about evolution theory that Albertans don't get? *Calgary Herald,* August 12. http://www.canada.com/calgaryherald/news/story.html?id=7cfcfff3-286a-4f29-9e81-e70951b54e4c.

Brem, S. K., Ranney, M., & Schindel, J. (2003). Perceived consequences of evolution: College students perceive negative personal and social impact in evolutionary theory. *Science Education, 87,* 181–206.

Bridewell, W., & Langley, P. (2010). Two kinds of knowledge in scientific discovery. *Topcs in Cognitive Science, 2,* 36–52.

Bridewell, W., Langley, P., Todorovski, L., & Dzeroski, S. (2008). Inductive process modeling. *Machine Learning, 71,* 1–32.

Bringsjord, S. (2008). Declarative/logic-based cognitive modeling. In R. Sun (Ed.), *The Cambridge handbook of computational psychology* (pp. 127–169). Cambridge: Cambridge University Press.

Brock, T. D. (1988). *Robert Koch: A life in medicine and bacteriology.* Madison, WI: Science Tech Publishers.

Brooks, F. P. (1982). *The mythical man-month: Essays on software engineering.* Reading, MA: Addison-Wesley.

Brown, J. R. (2001). *Who rules in science? An opinionated guide to the wars.* Cambridge, MA: Harvard University Press.

Brunaud, L., Alberto, J. M., Ayav, A., Gerard, P., Namour, F., Antunes, L., et al. (2003). Effects of vitamin B12 and folate deficiencies on DNA methylation and carcinogenesis in rat liver. *Clinical Chemistry and Laboratory Medicine, 41*(8), 1012–1019.

Buchwald, J. Z., & Smith, G. E. (1997). Thomas S. Kuhn, 1922–1996. *Philosophy of Science, 64*, 361–376.

Bunge, M. (2003). *Emergence and convergence: Qualitative novelty and the unity of knowledge.* Toronto: University of Toronto Press.

Bylander, T., Allemang, D., Tanner, M., & Josephson, J. (1991). The computational complexity of abduction. *Artificial Intelligence, 49*, 25–60.

Caldji, C., Diorio, J., & Meaney, M. J. (2000). Variations in maternal care in infancy regulate the development of stress reactivity. *Biological Psychiatry, 48*(12), 1164–1174.

Carey, S. (1985). *Conceptual change in childhood.* Cambridge, MA: MIT Press/Bradford Books.

Carey, S. (2009). *The origin of concepts.* Oxford: Oxford University Press.

Carruthers, P. (2011). Creative action in mind. *Philosophical Psychology, 24*(4), 437–461.

Carruthers, P., Stich, S., & Siegal, M. (Eds.). (2002). *The cognitive basis of science.* Cambridge: Cambridge University Press.

Chahrour, M., & Zoghbi, H. Y. (2007). The story of Rett syndrome: from clinic to neurobiology. *Neuron, 56*(3), 422–437.

Chalmers, D. J. (1996). *The conscious mind.* Oxford: Oxford University Press.

Champagne, F. A., & Curley, J. P. (2009). Epigenetic mechanisms mediating the long-term effects of maternal care on development. *Neuroscience and Biobehavioral Reviews, 33*(4), 593–600.

Champagne, F. A., Francis, D. D., Mar, A., & Meaney, M. J. (2003). Variations in maternal care in the rat as a mediating influence for the effects of environment on development. *Physiology & Behavior, 79*(3), 359–371.

Chandrasekharan, S. (2009). Building to discover: A common coding model. *Cognitive Science, 33*, 1059–1086.

Chemero, A. (2009). *Radical embodied cognitive science.* Cambridge, MA: MIT Press.

Cheng, P. W. (1985). Pictures of ghosts: A critique of Alfred Bloom's *The Linguistic Shaping of Thought. American Anthropologist, 87*, 917–922.

Chi, M. T. H. (2005). Commonsense conceptions of emergent processes: Why some misconceptions are robust. *Journal of the Learning Sciences, 14*, 161–199.

Chi, M. T. H. (2008). Three types of conceptual change: Belief revision, mental model transformation, and categorical shift. In S. Vosniadou (Ed.), *International handbook of research in conceptual change* (pp. 61–82). New York: Routledge.

Churchland, P. M. (1989). *A neurocomputational perspective*. Cambridge, MA: MIT Press.

Churchland, P. S., & Sejnowski, T. (1992). *The computational brain*. Cambridge, MA: MIT Press.

Clark, A. (1997). *Being there: Putting brain, body, and world together again*. Cambridge, MA: MIT Press.

Coelho, R. L. (2010). On the concept of force: How understanding its history can improve physics teaching. *Science & education, 19*, 91–113.

Comte, A. (1970). *Introduction to positive philosophy* (Ferré, F., Trans.). Indianapolis: Bobbs-Merrill.

Conklin, J., & Eliasmith, C. (2005). An attractor network model of path integration in the rat. *Journal of Computational Neuroscience, 18*, 183–203.

Cook, S. a. (1971). The complexity of theorem proving procedures. In *Proceedings of the Third Annual Symposium on Theory of Computing*, 151–158.

Cooney, C. A., Dave, A. A., & Wolff, G. L. (2002). Maternal methyl supplements in mice affect epigenetic variation and DNA methylation of offspring. *Journal of Nutrition, 132*(8 Suppl.), 2393S–2400S.

Cooper, B. (2005). Immigration and schizophrenia: The social causation hypothesis revisited. *British Journal of Psychiatry, 86*, 361–363.

Costello, F. J., & Keane, M. T. (2000). Efficient creativity: constraint-guided conceptual combination. *Cognitive Science, 24*, 299–349.

Craik, K. (1943). *The nature of explanation*. Cambridge: Cambridge University Press.

Craver, C. F. (2007). *Explaining the brain*. Oxford: Oxford University Press.

Crespi, B. (2008). Genomic imprinting in the development and evolution of psychotic spectrum conditions. *Biological Reviews of the Cambridge Philosophical Society, 83*(4), 441–493.

Crick, F. (1994). *The astonishing hypothesis: The scientific search for the soul*. London: Simon & Schuster.

Crowley, K., Schunn, C. D., & Okada, T. (Eds.). (2001). *Designing for science: Implications from everyday, classroom, and professional settings*. Mahwah, NJ: Erlbaum.

Csordas, A., Puschendorf, B., & Grunicke, H. (1986). Increased acetylation of histones at an early stage of oestradiol-mediated gene activation in the liver of immature chicks. *Journal of Steroid Biochemistry, 24*(1), 437–442.

Damasio, A. R. (1994). *Descartes' error: Emotion, reason, and the human brain*. New York: G. P. Putnam's Sons.

D'Andrade, R. (1995). *The development of cognitive anthropology.* Cambridge: Cambridge University Press.

Darden, L. (1983). Artificial intelligence and philosophy of science: Reasoning by analogy in theory construction. In P. Asquith & T. Nickles (Eds.), *PSA 1982* (Vol. 2, pp. 147–165). East Lansing.

Darden, L. (1991). *Theory change in science: Strategies from Mendelian genetics.* Oxford: Oxford University Press.

Darden, L. (2006). *Reasoning in biological discoveries.* Cambridge: Cambridge University Press.

Darden, L., & Cain, J. (1989). Selection type theories. *Philosophy of Science, 56,* 106–129.

Davidson, P. R., & Wolpert, D. M. (2005). Widespread access to predictive models in the motor system: A short review. *Journal of Neural Engineering, 2,* S313–S319.

Davies, W., Isles, A. R., & Wilkinson, L. S. (2005). Imprinted gene expression in the brain. *Neuroscience and Biobehavioral Reviews, 29*(3), 421–430.

Dawkins, R. (1976). *The selfish gene.* New York: Oxford University Press.

Dawkins, R. (2006). *The God delusion.* New York: Houghton Mifflin.

Dayan, P., & Abbott, L. F. (2001). *Theoretical neuroscience: Computational and mathematical modeling of neural systems.* Cambridge, MA: MIT Press.

Deci, E. L., & Ryan, R. M. (Eds.). (2002). *Handbook of self-determination research.* Rochester: Univerity of Rochester Press.

Delcuve, G. P., Rastegar, M., & Davie, J. R. (2009). Epigenetic control. *Journal of Cellular Physiology, 219*(2), 243–250.

Dembski, W. (1999). *Intelligent design: The bridge between science and theology.* Downers Grove, IL: InterVarsity Press.

Deniz, H., Donnelly, L. A., & Yilmaz, I. (2008). Exploring the factors related to acceptance of evolutionary theory among Turkish preservice biology teachers: Toward a more informative conceptual ecology for biological evolution. *Journal of Research in Science Teaching, 45,* 420–443.

Dennett, D. (2006). *Breaking the spell: Religion as a natural phenomenon.* New York: Penguin.

Descartes, R. (1985). *The philosophical writings of Descartes* (J. Cottingham et al., Trans.). Cambridge: Cambridge University Press.

diSessa, A. (1988). Knowledge in pieces. In G. Forman & P. Pufall (Eds.), *Constructivism in the computer age* (pp. 49–70). Hillsdale, NJ: Erlbaum.

Douglas, H. E. (2009). *Science, policy, and the value-free ideal*. Pittsburgh: University of Pittsburgh Press.

Dreyfus, H. L. (2007). Why Heideggerian AI failed and how fixing it would require making it more Heideggerian. *Philosophical Psychology, 20*, 247–268.

Dunbar, K. (1995). How scientists really reason: Scientific reasoning in real-world laboratories. In R. J. Sternberg & J. Davidson (Eds.), *Mechanisms of insight* (pp. 365–395). Cambridge, MA: MIT Press.

Dunbar, K. (1997). How scientists think: On-line creativity and conceptual change in science. In T. B. Ward, S. M. Smith, & J. Vaid (Eds.), *Creative thought: An investigation of conceptual structures and processes* (pp. 461–493). Washington: American Psychological Association.

Dunbar, K. (2001). What scientific thinking reveals about the nature of cognition. In K. Crowley, C. D. Schunn, & T. Okada (Eds.), *Designing for science: Implications from everyday, classroom, and professional settings* (pp. 115–140). Mahwah, NJ: Erlbaum.

Dunbar, K., & Fugelsang, J. (2005). Scientific thinking and reasoning. In K. J. Holyoak & R. Morrison (Eds.), *Cambridge handbook of thinking and reasoning* (pp. 705–726). Cambridge: Cambridge University Press.

Edelman, S. (2008). *Computing the mind: How the mind really works*. Oxford: Oxford University Press.

Eliasmith, C. (2004). Learning context sensitive logical inference in a neurobiological simulation. In S. Levy & R. Gayler (Eds.), *Compositional connectionism in cognitive science* (pp. 17–20). Menlo Park, CA: AAAI Press.

Eliasmith, C. (2005a). Cognition with neurons: A large-scale, biologically realistic model of the Wason task. In B. Bara, L. Barasalou & M. Bucciarelli (Eds.), *Proceedings of the XXVII Annual Conference of the Cognitive Science Society* (pp. 624–629). Mahwah, NJ: Erlbaum.

Eliasmith, C. (2005b). Neurosemantics and categories. In H. Cohen & C. Lefebvre (Eds.), *Handbook of categorization in cognitive science* (pp. 1035–1054). Amsterdam: Elsevier.

Eliasmith, C. (forthcoming). *How to build a brain*. Oxford: Oxford University Press.

Eliasmith, C., & Anderson, C. H. (2003). *Neural engineering: Computation, representation and dynamics in neurobiological systems*. Cambridge, MA: MIT Press.

Eliasmith, C., & Stewart, T. C. (forthcoming). A new biologically implemented cognitive architecture. Unpublished manuscript, University of Waterloo Centre for Theoretical Neuroscience.

Eliasmith, C., & Thagard, P. (1997). Waves, particles, and explanatory coherence. *British Journal for the Philosophy of Science, 48*, 1–19.

Eliasmith, C., & Thagard, P. (2001). Integrating structure and meaning: A distributed model of analogical mapping. *Cognitive Science, 25,* 245–286.

Ellis, J. H. (1987). The history of non-secret encryption. Retrieved Feb. 15, 2011, from http://cryptocellar.web.cern.ch/cryptocellar/cesg/ellis.pdf.

Elman, J. L. (1990). Finding structure in time. *Cognitive Science, 14,* 179–211.

Elman, J. L., Bates, E. A., Johnson, M. H., Karmiloff-Smith, A., Parisi, D., & Plunkett, K. (1996). *Rethinking innateness: A connectionist perspective on development.* Cambridge, MA: MIT Press.

Elster, J. (2007). *Explaining social behavior.* Cambridge: Cambridge University Press.

Engel, A. K., Fries, P., König, P., Brecht, M., & Singer, W. (1999). Temporal binding, binocular rivalry, and consciousness. *Consciousness and Cognition, 8,* 128–151.

Evans, E. M. (2008). Conceptual change and evolutionary biology: A developmental analysis. In S. Vosniadou (Ed.), *International handbook of research on conceptual change* (pp. 263–294). New York: Routledge.

Falkenhainer, B. (1990). A unified approach to explanation and theory formation. In J. Shrager & P. Langley (Eds.), *Computational models of discovery and theory formatio.* (pp. 157–196). San Mateo, CA: Morgan Kaufmann.

Falkenhainer, B., Forbus, K. D., & Gentner, D. (1989). The structure-mapping engine: Algorithms and examples. *Artificial Intelligence, 41,* 1–63.

Fauconnier, G., & Turner, M. (2002). *The way we think.* New York: Basic Books.

Feist, G. J. (2006). *The psychology of science and the origins of the scientific mind.* New Haven: Yale University Press.

Feyerabend, P. (1965). Problems of empiricism. In R. Colodny (Ed.), *Beyond the edge of certainty* (pp. 145–260). Pittsburgh: University of Pittsburgh Press.

Feynman, R. (1999). *The pleasure of finding things out.* Cambridge, MA: Perseus Books.

Findlay, S. D., & Thagard, P. (forthcoming). Emotional change in international negotiation: Analyzing the Camp David accords using cognitive-affective maps. *Group Decision and Negotiation.*

Finke, R., Ward, T. B., & Smith, S. M. (1992). *Creative cognition: Theory, research and applications.* Cambridge, MA: MIT Press/Bradford Books.

Flannery, T. (2006). *The weather makers.* Toronto: HarperCollins.

Fodor, J. (2000). *The mind doesn't work that way.* Cambridge, MA: MIT Press.

Forbus, K. D. (1984). Qualitative process theory. *Artificial Intelligence, 24,* 85–168.

Foucault, M. (1965). *Madness and civilization: A history of insanity in the age of reason.* New York: Pantheon.

Fraga, M. F., Ballestar, E., Paz, M. F., Ropero, S., Setien, F., Ballestar, M. L., et al. (2005). Epigenetic differences arise during the lifetime of monozygotic twins. *Proceedings of the National Academy of Sciences of the United States of America, 102*(30), 10604–10609.

Frenkel, K. A. (1987). Profiles in computing: Brian K. Reid: a graphics tale of a hacker tracker. *Communications of the ACM, 30,* 820–823.

Friedman, M., & Friedland, G. W. (1998). *Medicine's 10 greatest discoveries.* New Haven: Yale University Press.

Fuster, J. M. (2002). *Cortex and mind: Unifying cognition.* Oxford: Oxford University Press.

Gärdenfors, P. (1988). *Knowledge in flux.* Cambridge, MA: MIT Press/Bradford Books.

Gärdenfors, P. (Ed.). (1992). *Belief revision.* Cambridge: Cambridge University Press.

Gardner, H. (1985). *The mind's new science.* New York: Basic Books.

Gardner, M. (1978). *Aha! Insight.* New York: Scientific American/W. H. Freeman.

Garey, M., & Johnson, D. (1979). *Computers and intractability.* New York: Freeman.

Gentner, D., Brem, S., Ferguson, R., Wolff, P., Markman, A. B., & Forbus, K. (1997). Analogy and creativity in the works of Johannes Kepler. In T. B. Ward, S. M. Smith, & J. Vaid (Eds.), *Creative thought: An investigation of conceptual structures and processes* (pp. 403–459). Washington, DC: American Psychological Association.

Gentner, D., & Stevens, A. L. (Eds.). (1983). *Mental models.* Hillsdale, NJ: Erlbaum.

Georgopoulos, A. P., Schwartz, A. B., & Kettner, R. E. (1986). Neuronal population coding of movement direction. *Science, 233*(4771), 1416–1419.

Gholson, B., Shadish, W., Neimeyer, R., & Houts, A. (Eds.). (1989). *Psychology of science: Contributions to metascience.* Cambridge: Cambridge University Press.

Gibbs, R. W. (2006). *Embodiment and cognitive science.* Cambridge: Cambridge University Press.

Giere, R. N. (1987). The cognitive study of science. In N. Nersessian (Ed.), *The process of science* (pp. 139–160). Berlin: Springer.

Giere, R. N. (1988). *Explaining science: A cognitive approach.* Chicago: University of Chicago Press.

Giere, R. N. (Ed.). (1992). *Cognitive models of science* (Vol. 15). Minneapolis: University of Minnesota Press.

Giere, R. N. (1999). *Science without laws.* Chicago: University of Chicago Press.

Giere, R. N. (2002). Scientific cognition as distributed cognition. In P. Carruthers, S. Stich, & M. Seigal (Eds.), *The cognitive basis of science* (pp. 285–299). Cambridge: Cambridge University Press.

Giere, R. N. (2010). *Scientific perspectivism.* Chicago: University of Chicago Press.

Gilovich, T., Griffin, D., & Kahneman, D. (Eds.). (2002). *Heuristics and biases: The psychology of intuitive judgment.* Cambridge: Cambridge University Press.

Glymour, C. (2001). *The mind's arrows: Bayes nets and graphical causal models in psychology.* Cambridge, MA: MIT Press.

Glymour, C. (2003). Learning, prediction, and causal Bayes nets. *Trends in Cognitive Sciences, 7,* 43–48.

Goldman, A. (1986). *Epistemology and cognition.* Cambridge, MA: Harvard University Press.

Goldman, A. (1999). *Knowledge in a social world.* Oxford: Oxford University Press.

Goldstine, H. (1972). *The computer from Pascal to Von Neumann.* Princeton, NJ: Princeton University Press.

Gondhalekar, P. (2001). *The grip of gravity: The quest to understand the laws of motion and gravitation.* Cambridge: Cambridge University Press.

Gooding, D. (1990). *Experiment and the nature of meaning.* Dordrecht: Kluwer.

Goodwin, F. K., & Jamison, K. R. (2007). *Manic-depressive illness: Bipolar disorders and recurrent depression* (2nd ed.). Oxford: Oxford University Press.

Gopnik, A. (1998). Explanation as orgasm. *Minds and Machines, 8,* 101–118.

Gopnik, A., Glymour, C., Sobel, D. M., Schultz, L. E., Kushur, T., & Danks, D. (2004). A theory of causal learning in children: Causal maps and Bayes nets. *Psychological Review, 2004,* 3–32.

Gore, A. (2006). *An inconvenient truth.* Emmaus, PA: Rodale.

Gorman, M. E., Tweney, R. D., Gooding, D. C., & Kincannon, A. P. (Eds.). (2005). *Scientific and technological thinking.* Mahwah, NJ: Erlbaum.

Gottlob, G., Scarcello, F., & Sideri, M. (2002). Fixed-parameter complexity in AI and nonmonotonic reasoning. *Artificial Intelligence, 138,* 55–86.

Gottman, J. M., Tyson, R., Swanson, K. R., Swanson, C. C., & Murray, J. D. (2003). *The mathematics of marriage: Dynamic nonlinear models.* Cambridge, MA: MIT Press.

Gould, S. J. (1999). *Rock of ages: Science and religion in the fullness of life.* New York: Ballantine.

Grandjean, D., Sander, D., & Scherer, K. R. (2008). Conscious emotional experience emerges as a function of multilevel, appraisal-driven response synchronization. *Consciousness and Cognition, 17,* 484–495.

Grene, M., & Depew, D. (2004). *The philosophy of biology: An episodic history.* Cambridge: Cambridge University Press.

Griffiths, T. L., Kemp, C., & Tenenbaum, J. B. (2008). Bayesian models of cognition. In R. Sun (Ed.), *The Cambridge handbook of computational psychology* (pp. 59–100). Cambridge: Cambridge University Press.

Grisdale, C. D. W. (2010). *Conceptual change: Gods, elements, and water.* M.A. thesis, University of Waterloo, Waterloo.

Guo, A. (Ed.). (1992). *Huang di nei jing su wen jiao zhu* (Annotations on Yellow Emperor's classic of internal medicine: Plain questions; in Chinese). Beijing: People's Medical Press.

Hacking, I. (1975). *The emergence of probability.* Cambridge: Cambridge University Press.

Hadamard, J. (1945). *The psychology of invention in the mathematical field.* New York: Dover.

Hafner, K., & Lyon, M. (1996). *Where wizards stay up late: The origins of the internet.* New York: Simon & Schuster.

Hanson, N. R. (1958). *Patterns of discovery.* Cambridge: Cambridge University Press.

Hanson, N. R. (1965). Notes toward a logic of discovery. In R. J. Bernstein (Ed.), *Perspectives on peirce* (pp. 42–65). New Haven: Yale University Press.

Hardy-Vallée, B., & Thagard, P. (2008). How to play the ultimatum game: An engineering approach to metanormativity. *Philosophical Psychology, 21,* 173–192.

Harman, G. (1973). *Thought.* Princeton: Princeton University Press.

Harman, G. (1986). *Change in view: Principles of reasoning.* Cambridge, MA: MIT Press/ Bradford Books.

Harman, G. (1987). (Nonsolopsistic) conceptual role semantics. In E. LePore (Ed.), *Semantics of natural language* (pp. 55–81). New York: Academic Press.

Haven, K. (2007). *100 greatest science discoveries of all time.* Westport, CT: Libraries Unlimited.

Hebb, D. O. (1949). *The organization of behavior.* New York: Wiley.

Hebb, D. O. (1980). *Essay on mind.* Hillsdale, NJ: Erlbaum.

Hélie, S., & Sun, R. (2010). Incubation, insight, and creative problem solving: A unified theory and a connectionist model. *Psychological Review, 117*, 994–1024.

Hempel, C. G. (1965). *Aspects of scientific explanation.* New York: Free Press.

Henderson, D. K. (1994). Epistemic competence and contextualist epistemology: Why contextualism is not just the poor man's coherentism. *Journal of Philosophy, 91*, 627–649.

Hesse, M. (1966). *Models and analogies in science.* Notre Dame, IN: Notre Dame University Press.

Hiltzik, M. A. (1999). *Dealers of lightning: Xerox PARC and the dawn of the computer age.* New York: HarperCollins.

Hippocrates. (400 BC.). On the sacred disease. Retrieved Jan. 12, 2010, from http://classics.mit.edu/Hippocrates/sacred.html.

Hippocrates. (1988). *Hippocrates,* vol. V (P. Potter, Trans.). Cambridge, MA: Harvard University Press.

Hodges, A. (1983). *Alan Turing: The enigma.* London: Burnett Books.

Hofstadter, D. (1995). *Fluid concepts and creative analogies: Computer models of the fundamental mechanisms of thought.* New York: Basic Books.

Hokayem, H., & BouJaoude, S. (2008). Colllege students perceptions of the theory of evolution. *Journal of Research in Science Teaching, 45*, 395–419.

Holland, J. H., Holyoak, K. J., Nisbett, R. E., & Thagard, P. R. (1986). *Induction: Processes of inference, learning, and discovery.* Cambridge, MA: MIT Press/Bradford Books.

Holyoak, K. J., & Thagard, P. (1995). *Mental leaps: Analogy in creative thought.* Cambridge, MA: MIT Press/Bradford Books.

Holyoak, K. J., & Thagard, P. (1997). The analogical mind. *American Psychologist, 52*, 35–44.

Hooke, R. (1665). *Micrographia: Or some physiological descriptions of minute bodies made by magnifying glasses with observations and inquiries thereupon.* London: John Martin and James Allestry.

Hopfield, J. (1999). Odor space and olfactory processing: Collective algorithms and neural implementation. *Proceedings of the National Academy of Sciences of the United States of America, 96*, 12506–12511.

Horwich, P. (Ed.). (1993). *World changes: Thomas Kuhn and the nature of science.* Cambridge, MA: MIT Press.

Human-Genome-Project. (2006). Human genome project information. http://www.ornl.gov/sci/techresources/Human_Genome/home.shtml.

Hume, D. (1888). *A treatise of human nature.* Oxford: Clarendon Press.

Hummel, J. E., & Holyoak, K. J. (2003). A symbolic-connectionist theory of relational inference and generalization. *Psychological Review, 110,* 220–264.

Hummel, J. E., & Holyoak, K. J. (1997). Distributed representations of structure: A theory of analogical access and mapping. *Psychological Review, 104,* 427–466.

Isen, A. M. (1993). Positive affect and decision making. In M. Lewis & J. M. Haviland (Eds.), *Handbook of emotions* (pp. 261–277). New York: Guilford Press.

Jacobson, L., & Sapolsky, R. (1991). The role of the hippocampus in feedback regulation of the hypothalamic-pituitary-adrenocortical axis. *Endocrine Reviews, 12*(2), 118–134.

James, W. (1884). What is an emotion? *Mind, 9,* 188–205.

Jammer, M. (1957). *Concepts of force.* Cambridge, MA: Harvard University Press.

Jirtle, R. L., & Skinner, M. K. (2007). Environmental epigenomics and disease susceptibility. *Nature Reviews. Genetics, 8*(4), 253–262.

Johnson, J. S., Spencer, J. P., & Schöner, G. (2008). Moving to higher ground: The dynamic field theory and the dynamics of visual cognition. *New Ideas in Psychology, 26,* 227–251.

Johnson-Laird, P. N. (1983). *Mental models.* Cambridge, MA: Harvard University Press.

Johnson-Laird, P. N. (2004). The history of mental models. In K. Manktelow & M. C. Chung (Eds.), *Psychology of reasoning: Theoretical and historical perspectives* (pp. 179–212). New York: Psychology Press.

Johnson-Laird, P. N. (2006). *How we reason.* Oxford: Oxford University Press.

Johnson-Laird, P. N., & Byrne, R. M. (1991). *Deduction.* Hillsdale, NJ: Lawrence Erlbaum Associates.

Josephson, J. R., & Josephson, S. G. (Eds.). (1994). *Abductive inference: Computation, philosophy, technology.* Cambridge: Cambridge University Press.

Kaas, J. H. (1997). Topographic maps are fundamental to sensory processing. *Brain Research Bulletin, 44,* 107–112.

Kahneman, D., Slovic, P., & Tversky, A. (1982). *Judgment under uncertainty: Heuristics and biases.* New York: Cambridge University Press.

Kampourakis, K., & Zogza, V. (2008). Students' intuitive explanations of the causes of homologies and adaptations. *Science & Education, 17,* 27–47.

Kaufman, J. C., & Baer, J. (Eds.). (2005). *Creativity across domains: Faces of the muse.* Mahwah, NJ: Erlbaum.

Kay, A. C. (1996a). The early history of SmallTalk. In T. J. Bergin & R. G. Gibson (Eds.), *History of programming languages II* (pp. 511–579). New York: Addison-Wesley.

Kay, A. C. (1996b). Transcript of SmallTalk presentation. In T. J. Bergin & R. G. Gibson (Eds.), *History of programming languages II*. New York: Addison-Wesley.

Ke, X., Lei, Q., James, S. J., Kelleher, S. L., Melnyk, S., Jernigan, S., et al. (2006). Uteroplacental insufficiency affects epigenetic determinants of chromatin structure in brains of neonatal and juvenile IUGR rats. *Physiological Genomics, 25*(1), 16–28.

Keil, F. C. (2006). Explanation and understanding. *Annual Review of Psychology, 57,* 227–254.

Keil, F. C., & Wilson, R. A. (Eds.). (2000). *Explanation and cognition.* Cambridge, MA: MIT Press.

Kelley, H. H. (1973). The process of causal attribution. *American Psychologist, 28,* 107–128.

Kertesz, A. (2004). *Cognitive semantics and scientific knowledge: Case studies in the cognitive science of science.* Amsterdam: John Benjamins.

King, R. D., Rowland, J., Oliver, S. G., Young, M., et al. (2009). The automation of science. *Science, 324,* 85–89.

Kitayama, S., & Cohen, D. (Eds.). (2007). *Handbook of cultural psychology.* New York: Guilford Press.

Kitcher, P. (1981). Explanatory unification. *Philosophy of Science, 48,* 507–531.

Kitcher, P. (1993). *The advancement of science.* Oxford: Oxford University Press.

Kitcher, P. (2001). Real realism: The Galilean strategy. *Philosophical Review, 110,* 151–197.

Kitcher, P. (2002). On the explanatory role of correspondence truth. *Philosophy and Phenomenological Research, 64,* 346–364.

Kitcher, P., & Salmon, W. (1989). *Scientific explanation.* Minneapolis: University of Minnesota Press.

Klahr, D. (2000). *Exploring science: The cognition and development of discovery processes.* Cambridge, MA: MIT Press.

Knudsen, E. I., du Lac, S., & Esterly, S. D. (1987). Computational maps in the brain. *Annual Review of Neuroscience, 10,* 41–65.

Knuth, D. E. (1974). *Computer programming as an art ACM Turing Award lectures: The first twenty years, 1966–1985.* New York: ACM Press.

Koestler, A. (1967). *The act of creation.* New York: Dell.

Konolige, K. (1992). Abduction versus closure in causal theories. *Artificial Intelligence, 53,* 255–272.

Kounios, J., & Beeman, M. (2009). The *Aha!* moment: The cognitive neuroscience of insight. *Current Directions in Psychological Science, 18,* 210–216.

Koza, J. R. (1992). *Genetic programming.* Cambridge, MA: MIT Press.

Kuhn, T. S. (1962). *The structure of scientific revolutions.* Chicago: University of Chicago Press.

Kuhn, T. S. (1970). *The structure of scientific revolutions* (2nd ed.). Chicago: University of Chicago Press.

Kuhn, T. S. (1993). Afterwords. In P. Horwich (Ed.), *World changes: Thomas Kuhn and the nature of science* (pp. 311–341). Cambridge, MA: MIT Press.

Kuipers, T. (2000). *From instrumentalism to constructive realism.* Dordrecht: Kluwer.

Kulkarni, D., & Simon, H. (1988). The processes of scientific discovery: The strategy of experimentation. *Cognitive Science, 12,* 139–175.

Kulkarni, D., & Simon, H. (1990). Experimentation in scientific discovery. In J. Shrager & P. Langley (Eds.), *Computational models of discovery and theory formation* (pp. 255–273). San Mateo, CA: Morgan Kaufmann.

Kunda, Z. (1990). The case for motivated reasoning. *Psychological Bulletin, 108,* 480–498.

Kunda, Z. (1999). *Social cognition: Making sense of people.* Cambridge, MA: MIT Press.

Kunda, Z., Miller, D., & Claire, T. (1990). Combining social concepts: The role of causal reasoning. *Cognitive Science, 14,* 551–577.

Kutchins, H., & Kirk, S. (1997). *Making us crazy.* New York: Free Press.

Lacey, H. (1999). *Is science value free? Values and scientific understanding.* London: Routledge.

Laing, R. D. (1967). *The politics of experience.* New York: Pantheon.

Laird, J., Rosenbloom, P., & Newell, A. (1986). Chunking in Soar: The anatomy of a general learning mechanism. *Machine Learning, 1,* 11–46.

Lamb, J., Crawford, E. D., Peck, D., Modell, J. W., Blat, I. C., Wrobel, M. J., et al. (2006). The Connectivity Map: Using gene-expression signatures to connect small molecules, genes, and disease. *Science, 313*(5795), 1929–1935.

Langley, P. (1996). *Elements of machine learning.* San Francisco: Morgan Kaufmann.

Langley, P., Simon, H., Bradshaw, G., & Zytkow, J. (1987). *Scientific discovery.* Cambridge, MA: MIT Press/Bradford Books.

LaPorte, J. (2004). *Natural kinds and conceptual change.* Cambridge: Cambridge University Press.

LaSalle, J. M., Hogart, A., & Thatcher, K. N. (2005). Rett syndrome: a Rosetta stone for understanding the molecular pathogenesis of autism. *International Review of Neurobiology, 71,* 131–165.

Latour, B. (2004). Why has critique run out of steam? From matters of fact to matters of concern. *Critical Inquiry, 30,* 225–247.

Latour, B., & Woolgar, S. (1986). *Laboratory life: The construction of scientific facts.* Princeton, NJ: Princeton University Press.

Laudan, L. (1981a). A confutation of convergent realism. *Philosophy of Science, 48,* 19–49.

Laudan, L. (1981b). *Science and hypothesis.* Dordrecht: Reidel.

Laudan, L. (1984). *Science and values.* Berkeley: University of California Press.

Laudan, L. (1990). *Science and relativism.* Chicago: University of Chicago Press.

Leake, D. B. (1992). *Evaluating explanations: A content theory.* Hillsdale, NJ: Erlbaum.

Lehrer, K. (1990). *Theory of knowledge.* Boulder: Westview.

Lenat, D., & Brown, J. S. (1984). Why AM and Eurisko appear to work. *Artificial Intelligence, 23,* 269–294.

Levenson, J. M., & Sweatt, J. D. (2005). Epigenetic mechanisms in memory formation. *Nature Reviews. Neuroscience, 6,* 108–118.

Levin, M. (1984). What kind of explanation is truth? In J. Leplin (Ed.), *Scientific realism* (pp. 124–139). Berkeley: University of California Press.

Levy, S. (1984). *Hackers: Heroes of the computer revolution.* Garden City, N.Y.: Anchor Press/Doubleday.

Lewis, D. (1986). *Philosophical papers.* Oxford: Oxford University Press.

Lindsay, R., Buchanan, B., Feigenbaum, E., & Lederberg, J. (1980). *Applications of organic chemistry for organic chemistry: The DENDRAL project.* New York: McGraw Hill.

Lipson, H., & Pollock, J. B. (2000). Automatic design and manufacture of robotic lifeforms. *Nature, 406,* 974–978.

Lipton, P. (2004). *Inference to the best explanation* (2nd ed.). London: Routledge.

Litt, A., Eliasmith, C., & Thagard, P. (2008). Neural affective decision theory: Choices, brains, and emotions. *Cognitive Systems Research, 9,* 252–273.

Lloyd, G. E. R. (1996). *Adversaries and authorities*. Cambridge: Cambridge University Press.

Lodish, H., Berk, A., Zipursky, S. L., Matsudaira, P., Baltimore, D., & Darnell, J. (2000). *Molecular cell biology* (4th ed.). New York: W. H. Freeman.

Lomborg, B. (2007). *Cool it: The skeptical environmentalist's guide to global warming*. Toronto: Random House.

Lombrozo, T. (2009). Explanation and categorization: How "why?" informs "what?." *Cognition*, *110*, 248–253.

Longino, H. (1990). *Science as social knowledge: Values and objectivity in scientific inquiry*. Princeton: Princeton University Press.

Lu, G., & Needham, J. (1980). *Celestial lancets: A history and rationale of acupuncture and moxa*. Cambridge: Cambridge University Press.

Machamer, P., Darden, L., & Craver, C. F. (2000). Thinking about mechanisms. *Philosophy of Science*, *67*, 1–25.

Machery, E. (2009). *Doing without concepts*. Oxford: Oxford University Press.

Mackintosh, A. R. (1988). Dr. Atanasoff's computer. Retrieved Feb. 15, 2011, from http://www.webcitation.org/query?id=1257001829419648.

MacLennan, N. K., James, S. J., Melnyk, S., Piroozi, A., Jernigan, S., Hsu, J. L., et al. (2004). Uteroplacental insufficiency alters DNA methylation, one-carbon metabolism, and histone acetylation in IUGR rats. *Physiological Genomics*, *18*(1), 43–50.

Magnani, L. (1999). Model-based creative abduction. In L. Magnani, P. Nersessian, & P. Thagard (Eds.), *Model-based reasoning in scientific discovery* (pp. 219–238). New York: Kluwer.

Magnani, L. (2001). *Abduction, reason, and science: Processes of discovery and explanation*. New York: Kluwer/Plenum.

Magnani, L. (2009). *Abductive cognition: The epistemological and eco-cognitive dimensions of hypothetical reasoning*. Berlin: Springer.

Mandler, J. M. (2004). *The foundations of mind: Origins of conceptual thought*. Oxford: Oxford University Press.

Mayr, E. (1982). *The growth of biological thought*. Cambridge, MA: Harvard University Press.

McCormick, D. A., Connors, B. W., Lighthall, J. W., & Prince, D. A. (1985). Comparative electrophysiology of pyramidal and sparsely spiny stellate neurons of the neocortex. *Journal of Neurophysiology*, *54*, 782–806.

Medin, D., & Shoben, E. (1988). Context and structure in conceptual combination. *Cognitive Psychology, 20,* 158–190.

Medin, D. L. (1989). Concepts and conceptual structure. *American Psychologist, 44,* 1469–1481.

Mednick, S. A. (1962). The associate basis of the creative process. *Psychological Review, 69,* 220–232.

Meyer, J. S., & Quenzer, L. F. (2005). *Psychopharmacology: Drugs, the brain, and behavior.* Sunderland, MA: Sinauer.

Meyers, M. A. (2007). *Happy accidents: Serendipity in modern medical breakthroughs.* New York: Arcade Publishing.

Michotte, A. (1963). *The perception of causality* (Miles, T. R., & Miles, E., Trans.). London: Methuen.

Mill, J., & Petronis, A. (2007). Molecular studies of major depressive disorder: the epigenetic perspective. *Molecular Psychiatry, 12*(9), 799–814.

Mill, J. S. (1970). *A system of logic* (8th ed.). London: Longman.

Mill, J. S. (1974). *A system of logic ratiocinative and inductive.* Toronto: University of Toronto Press.

Miller, G., & Johnson-Laird, P. (1976). *Language and perception.* Cambridge, MA: Harvard University Press.

Miller, J. D., Scott, E. C., & Okamoto, S. (2006). Science communication. Public acceptance of evolution. *Science, 313*(5788), 765–766.

Millgram, E. (2000). Coherence: The price of the ticket. *Journal of Philosophy, 97,* 82–93.

Minsky, M. (1975). A framework for representing knowledge. In P. H. Winston (Ed.), *The psychology of computer vision* (pp. 211–277). New York: McGraw-Hill.

Minton, S., Carbonell, J. G., Knoblock, C. A., Kuokka, D. R., Etzioni, O., & Gil, Y. (1989). Explanation-based learning: A problem solving perspective. *Artificial Intelligence, 40,* 63–118.

Moyers, B. (1993). *Healing and the mind.* New York: Doubleday.

Mumford, M. D. (2002). Social innovation: Ten cases from Benjamin Franklin. *Creativity Research Journal, 14,* 253–266.

Murphy, D. (2006). *Psychiatry in the scientific image.* Cambridge, MA: MIT Press.

Murphy, G. L. (2002). *The big book of concepts.* Cambridge, MA: MIT Press.

Murray, C. J., & Lopez, A. D. (1997). Global mortality, disability, and the contribution of risk factors: Global Burden of Disease Study. *Lancet, 349*(9063), 1436–1442.

Nagarajan, R. P., Hogart, A. R., Gwye, Y., Martin, M. R., & LaSalle, J. M. (2006). Reduced MeCP2 expression is frequent in autism frontal cortex and correlates with aberrant MECP2 promoter methylation. *Epigenetics; Official Journal of the DNA Methylation Society, 1*(4), e1–e11.

Neapolitan, R. (1990). *Probabilistic reasoning in expert systems.* New York: John Wiley.

Nersessian, N. (1984). *Faraday to Einstein: Constructing meaning in scientific theories.* Dordrecht: Martinus Nijhoff.

Nersessian, N. (1992). How do scientists think? Capturing the dynamics of conceptual change in science. In R. Giere (Ed.), *Cognitive models of science* (Vol. 15, pp. 3–44). Minneapolis: University of Minnesota Press.

Nersessian, N. (2008). *Creating scientific concepts.* Cambridge, MA: MIT Press.

Nersessian, N. (2009). How do engineering scientists think? Model-based simulation in biomedical engineering laboratories. *Topics in Cognitive Science, 1,* 730–757.

Newell, A. (1990). *Unified theories of cognition.* Cambridge, MA: Harvard University Press.

Newell, A., Shaw, J. C., & Simon, H. (1958). Elements of a theory of human problem solving. *Psychological Review, 65,* 151–166.

Newell, A., & Simon, H. A. (1972). *Human problem solving.* Englewood Cliffs, NJ: Prentice-Hall.

Newton-Smith, W. H. (1981). *The rationality of science.* London: Routledge & Kegan Paul.

Nickles, T. (Ed.). (1980). *Scientific discovery, logic, and rationality.* Dordrecht: Reidel.

NIH. (1997). Acupuncture. NIH Consensus Statement Online , Nov. 3–5. Retrieved Sept. 6, 2011, from http://consensus.nih.gov/1997/1997Acupuncture107html.htm.

Nikitina, T., Ghosh, R. P., Horowitz-Scherer, R. A., Hansen, J. C., Grigoryev, S. A., & Woodcock, C. L. (2007). MeCP2-chromatin interactions include the formation of chromatosome-like structures and are altered in mutations causing Rett syndrome. *Journal of Biological Chemistry, 282*(38), 28237–28245.

Nikitina, T., Shi, X., Ghosh, R. P., Horowitz-Scherer, R. A., Hansen, J. C., & Woodcock, C. L. (2007). Multiple modes of interaction between the methylated DNA binding protein MeCP2 and chromatin. *Molecular and Cellular Biology, 27*(3), 864–877.

Nowak, G., & Thagard, P. (1992a). Copernicus, Ptolemy, and explanatory coherence. In R. Giere (Ed.), *Cognitive models of science* (Vol. 15, pp. 274–309). Minneapolis: University of Minnesota Press.

Nowak, G., & Thagard, P. (1992b). Newton, Descartes, and explanatory coherence. In R. Duschl & R. Hamilton (Eds.), *Philosophy of Science, Cognitive Psychology and Educational Theory and Practice* (pp. 69–115). Albany: SUNY Press.

Ohlsson, S. (2011). *Deep learning: How the mind overrides experience.* Cambridge: Cambridge University Press.

Olsson, E. (2005). *Against coherence: Truth, probability, and justification.* Oxford: Oxford University Press.

Olsson, E. J. (2002). What is the problem of coherence and truth? *Journal of Philosophy, 99,* 246–272.

O'Reilly, R. C., & Munakata, Y. (2000). *Computational explorations in cognitive neuroscience.* Cambridge, MA: MIT Press.

Osbeck, L. M., Nersessian, N. J., Malone, K. R., & Newstetter, W. C. (2011). *Science as psychology: Sense-making and identity in scientific practice.* Cambridge: Cambridge University Press.

Paley, W. (1963). *Natural theology: Selections.* Indianapolis: Bobbs-Merrill.

Parisien, C., & Thagard, P. (2008). Robosemantics: How Stanley the Volkswagen represents the world. *Minds and Machines, 18,* 169–178.

Partington, J. (1961). *A history of chemistry.* London: Macmillan.

Pasqualini, J. R., Mercat, P., & Giambiagi, N. (1989). Histone acetylation decreased by estradiol in the MCF-7 human mammary cancer cell line. *Breast Cancer Research and Treatment, 14*(1), 101–105.

Pearl, J. (1988). *Probabilistic reasoning in intelligent systems.* San Mateo: Morgan Kaufmann.

Pearl, J. (2000). *Causality: Models, reasoning, and inference.* Cambridge: Cambridge University Press.

Peirce, C. S. (1931–1958). *Collected papers.* Cambridge, MA: Harvard University Press.

Peirce, C. S. (1992). *Reasoning and the logic of things.* Cambridge, MA: Harvard University Press.

Pereira, F. C. (2007). *Creativity and artificial intelligence.* Berlin: Mouton de Gruyter.

Perkins, D. (2001). *The Eureka effect: The art and logic of breakthrough thinking.* New York: Norton.

Philbin, T. (2003). *The 100 greatest inventions of all time.* New York: Citadel Press.

Plate, T. (2003). *Holographic reduced representations.* Stanford: CSLI Publications.

Pogribny, I. P., Miller, B. J., & James, S. J. (1997). Alterations in hepatic p53 gene methylation patterns during tumor progression with folate/methyl deficiency in the rat. *Cancer Letters, 115*(1), 31–38.

Poincaré, H. (1913). *The foundations of science* (G. Halsted, Trans.). New York: Science Press.

Popper, K. (1959). *The logic of scientific discovery*. London: Hutchinson.

Popper, K. (1978). Natural selection and the emergence of mind. *Dialectica, 32*, 339–355.

Porkert, M., & Ullmann, C. (1988). *Chinese medicine* (M. Howson, Trans. 1st U.S. ed.). New York: Morrow.

Porter, R. (2002). *Madness: A brief history*. Oxford: Oxford University Press.

Poulter, M. O., Du, L., Weaver, I. C., Palkovits, M., Faludi, G., Merali, Z., et al. (2008). GABAA receptor promoter hypermethylation in suicide brain: implications for the involvement of epigenetic processes. *Biological Psychiatry, 64*(8), 645–652.

Preston, J., & Epley, N. (2009). Science and God: An automatic opposition between ultimate explanations. *Journal of Experimental Social Psychology, 45*, 238–241.

Prinz, J. (2004). *Gut reactions: A perceptual theory of emotion*. Oxford: Oxford University Press.

Proctor, R., & Capaldi, E. J. (Eds.). (forthcoming). *Psychology of science*. Oxford: Oxford University Press.

Psillos, S. (1999). *Scientific realism: How science tracks the truth*. London: Routledge.

Putnam, H. (1975). *Mind, language, and reality*. Cambridge: Cambridge University Press.

Putnam, H. (1983). There is at least one a priori truth. In H. Putnam (Ed.), *Realism and reason: Philosophical papers* (Vol. 3, pp. 98–114). Cambridge: Cambridge University Press.

Quartz, S. R., & Sejnowski, T. J. (1997). The neural basis of cognitive development: a constructivist manifesto. *Behavioral and Brain Sciences, 20*, 537–556.

Quartz, S. R., & Sejnowski, T. J. (2002). *Liars, lovers, and heroes: What the new brain science reveals about how we become who we are*. New York: William Morrow.

Quine, W. V. O. (1968). Epistemology naturalized. In W. V. O. Quine, *Ontological relativity and other essays* (pp. 69–90). New York: Columbia University Press.

Ranney, M., & Thanukos, A. (2010). Accepting evolution or creation in people, critters, plants, and classrooms: The maelstrom of American cognition about biological change. In R. Taylor & M. Ferrari (Eds.), *Evolution, epistemology, and science education* (pp. 143–172). Milton Park: Routledge.

Raymond, E. S. (1991). *The new hacker's dictionary*. Cambridge, MA: MIT Press.

Read, S., & Marcus-Newhall, A. (1993). The role of explanatory coherence in the construction of social explanations. *Journal of Personality and Social Psychology, 65,* 429–447.

Reichenbach, H. (1938). *Experience and prediction*. Chicago: University of Chicago Press.

Rescher, N. (1973). *The coherence theory of truth*. Oxford: Clarendon Press.

Richards, E. J. (2006). Inherited epigenetic variation—revisiting soft inheritance. *Nature Reviews. Genetics, 7*(5), 395–401.

Richardson, R. C. (2007). *Evolutionary psychology as maladapted psychology*. Cambridge, MA: MIT Press.

Rips, L. J. (1986). Mental muddles. In M. Brand & R. M. Harnish (Eds.), *The representation of knowledge and belief* (pp. 258–286). Tucson: University of Arizona Press.

Roberts, R. M. (1989). *Serendipity: Accidental discoveries in science*. New York: Wiley.

Roskies, A. L. (1999). The binding problem. *Neuron, 24,* 7–9.

Rott, H. (2000). Two dogmas of belief revision. *Journal of Philosophy, 97,* 503–522.

Rudner, R. (1961). Value judgments in the acceptance of theories. In P. G. Frank (Ed.), *The validation of scientific theories* (pp. 31–35). New York: Collier Books.

Rumelhart, D. E., & McClelland, J. L. (Eds.). (1986). *Parallel distributed processing: Explorations in the microstructure of cognition*. Cambridge, MA: MIT Press/Bradford Books.

Russell, S., & Norvig, P. (2003). *Artificial intelligence: A modern approach* (2nd ed.). Upper Saddle River, NJ: Prentice Hall.

Sackett, D. L., Rosenberg, W. M. C., Gray, J. A. M., Haynes, R. B., & Richardson, W. S. (1996). Evidence-based medicine: What it is and what it isn't. *British Medical Journal, 312,* 71–72.

Sahdra, B., & Thagard, P. (2003). Procedural knowledge in molecular biology. *Philosophical Psychology, 16,* 477–498.

Salinas, E., & Abbott, L. F. (1994). Vector reconstruction from firing rates. *Journal of Computational Neuroscience, 1,* 89–107.

Salmon, W. (1970). Statistical explanation. In R. Colodny (Ed.), *The nature and function of scientific theories* (pp. 173–231). Pittsburgh: University of Pittsburgh Press.

Salmon, W. (1984). *Scientific explanation and the causal structure of the world*. Princeton: Princeton University Press.

Salmon, W. C. (1989). Four decades of scientific explanation. In P. Kitcher & W. C. Salmon (Eds.), *Scientific explanation* (Minnesota Studies in the Philosophy of Science, vol. XIII) (pp. 3–219). Minneapolis: University of Minnesota Press.

Samaco, R. C., Hogart, A., & LaSalle, J. M. (2005). Epigenetic overlap in autism-spectrum neurodevelopmental disorders: MECP2 deficiency causes reduced expression of UBE3A and GABRB3. *Human Molecular Genetics, 14*(4), 483–492.

Sander, D., Grandjean, D., & Scherer, K. R. (2005). A systems approach to appraisal mechanisms in emotion. *Neural Networks, 18*, 317–352.

Schaffner, K. F. (1993). *Discovery and explanation in biology and medicine.* Chicago: University of Chicago Press.

Schank, P., & Ranney, M. (1991). Modeling an experimental study of explanatory coherence. In *Proceedings of the Thirteenth Annual Conference of the Cognitive Science Society* (pp. 892–897). Hillsdale, NJ: Erlbaum.

Schank, P., & Ranney, M. (1992). Assessing explanatory coherence: A new method for integrating verbal data with models of on-line belief revision. In *Proceedings of the Fourteenth Annual Conference of the Cognitive Science Society* (pp. 599–604). Hillsdale, NJ: Erlbaum.

Schank, R. C. (1986). *Explanation patterns: Understanding mechanically and creatively.* Hillsdale, NJ: Erlbaum.

Schroeder, F. A., Lin, C. L., Crusio, W. E., & Akbarian, S. (2007). Antidepressant-like effects of the histone deacetylase inhibitor, sodium butyrate, in the mouse. *Biological Psychiatry, 62*(1), 55–64.

Schunn, C. D., & Anderson, J. R. (1999). The generality/specificity of expertise in scientific reasoning. *Cognitive Science, 23*, 337–370.

Science-Channel. (2006). 100 greatest discoveries: Medicine. Retrieved Oct. 4, 2006, from http://science.discovery.com/convergence/100discoveries/big100/medicine.html.

Shapin, S. (1994). *A social history of truth.* Chicago: University of Chicago Press.

Shasha, D. (1995). *Out of their minds: The lives and discoveries of 15 great computer scientists.* New York: Copernicus.

Shastri, L., & Ajjanagadde, V. (1993). From simple associations to systematic reasoning: A connectionist representation of rules, variables, and dynamic bindings. *Behavioral and Brain Sciences, 16*, 417–494.

Shelley, C. (2003). *Multiple analogies in science and philosophy.* Amsterdam: John Benjamins.

Shelley, C. P. (1996). Visual abductive reasoning in archaeology. *Philosophy of Science, 63*, 278–301.

Shen, Z., & Chen, Z. (1994). *The basis of traditional Chinese medicine*. Boston: Shambhala.

Shogenji, T. (1999). Is coherence truth-conducive? *Analysis, 61*, 338–345.

Shorter, E. (1997). *A history of psychiatry*. New York: John Wiley & Sons.

Shrager, J., & Langley, P. (Eds.). (1990). *Computational models of scientific discovery and theory formation*. San Mateo: Morgan Kaufmann.

Shtulman, A. (2006). Qualitative differences between naive and scientific theories of evolution. *Cognitive Psychology, 52*(2), 170–194.

Shtulman, A., & Schulz, L. (2008). The relation between essentialist beliefs and evolutionary reasoning. *Cognitive Science, 32*, 1049–1062.

Simon, H. A. (1966). Scientific discovery and the psychology of problem solving. In R. G. Colodny (Ed.), *Mind and cosmos: Essays in contemporary science and philosophy* (pp. 22–40). Pittsburgh: University of Pittsburgh Press.

Simonton, D. (1988). *Scientific genius: A psychology of science*. Cambridge: Cambridge University Press.

Simonton, D. K. (2010). Creative thought as blind-variation and selective-retention: Combinatorial models of exceptional creativity. *Physics of Life Reviews, 7*, 156–179.

Sinatra, G. M., & Pintrich, P. R. (Eds.). (2003). *Intentional conceptual change*. Mahwah: Erlbaum.

Sinatra, G. M., Southerland, S. A., McConaughy, F., & Demastes, J. W. (2003). Intentions and beliefs in students understanding and acceptance of biological evolution. *Journal of Research in Science Teaching, 40*, 519–528.

Sinclair, K. D., Allegrucci, C., Singh, R., Gardner, D. S., Sebastian, S., Bispham, J., et al. (2007). DNA methylation, insulin resistance, and blood pressure in offspring determined by maternal periconceptional B vitamin and methionine status. *Proceedings of the National Academy of Sciences of the United States of America, 104*(49), 19351–19356.

Slater, A., & Quinn, P. C. (2001). Face recognition in the newborn infant. *Infant and Child Development, 10*, 21–24.

Slater, R. (1987). *Portraits in silicon*. Cambridge, MA: MIT Press.

Smith, E., & Osherson, D. (1984). Conceptual combination with prototype concepts. *Cognitive Science, 8*, 337–361.

Smith, E. E., & Kosslyn, S. M. (2007). *Cognitive psychology: Mind and brain*. Upper Saddle River, NJ: Pearson Prentice Hall.

Smolensky, P. (1990). Tensor product variable binding and the representation of symbolic structures in connectionist systems. *Artificial Intelligence, 46*, 159–217.

Smolin, L. (2001). *Three roads to quantum gravity.* New York: Basic Books.

Sober, E. (2008). *Evidence and evolution: The logic behind the science.* Cambridge: Cambridge University Press.

Sober, E., & Wilson, D. S. (1998). *Unto others: The evolution and psychology of unselfish behavior.* Cambridge, MA: Harvard University Press.

Spirtes, P., Glymour, C., & Scheines, R. (1993). *Causation, prediction, and search.* New York: Springer-Verlag.

Stanford, K. S. (2003). Pyrrhic victories for scientific realism. *Journal of Philosophy, 100,* 553–573.

Steinhardt, P. J., & Turok, N. (2007). *Endless universe: Beyond the big bang.* New York: Doubleday.

Sternberg, R. J., & Davidson, J. E. (Eds.). (1995). *The nature of insight.* Cambridge, MA: MIT Press.

Stewart, T. C., Choo, X., & Eliasmith, C. (2010a). Dynamic behaviour of a spiking model of action selection in the basal ganglia. In D. Salvucci & G. Gunzelmann (Eds.), *Proceedings of the 10th International Conference on Cognitive Modeling* (pp. 235–240). Philadelphia, PA: Drexel University Press.

Stewart, T. C., Choo, X., & Eliasmith, C. (2010b). Symbolic reasoning in spiking neurons: A model of the cortex/basal ganglia/thalamus loop. In S. Ohlsson & R. Catrambone (Eds.), *Proceedings of the 32nd Annual Conference of the Cognitive Science Society* (pp. 1100–1105). Portland, OR: Cognitive Science Society.

Stewart, T. C., & Eliasmith, C. (2009a). Compositionality and biologically plausible models. In W. Hinzen, E. Machery, & M. Werning (Eds.), *Oxford handbook of compositionality.* Oxford: Oxford University Press.

Stewart, T. C., & Eliasmith, C. (2009b). Spiking neurons and central executive control: The origin of the 50-millisecond cognitive cycle. In A. Howes, D. Peebles & R. Cooper (Eds.), *Proceedings of the 9th International Conference on Cognitive Modeling.* Manchester.

Straus, E. W., & Straus, A. (2006). *Medical marvels: The 100 greatest advances in medicine.* Buffalo: Prometheus Books.

Subranamiam, K., Kounios, J., Parrish, T. B., & Jung-Beeman, M. (2009). A brain mechanism for facilitation of insight by positive affect. *Journal of Cognitive Neuroscience, 21,* 415–432.

Sulloway, F. J. (1996). *Born to rebel: Birth order, family dynamics, and creative lives* (1st ed.). New York: Pantheon Books.

Sun, R. (2008a). Cognitive social stimulation. In R. Sun (Ed.), *Cambridge handbook of computational psychology* (pp. 530–548). Cambridge: Cambridge University Press.

Sun, R. (Ed.). (2008b). *The Cambridge handbook of computational psychology*. Cambridge: Cambridge University Press.

Sun, R., Coward, L. A., & Zenzen, M. J. (2005). On levels of cognitive modeling. *Philosophical Psychology, 18*, 613–637.

Sweeny, A. (2009). *BlackBerry planet*. Missisaugua, ON: Wiley.

Szasz, T. S. (1961). *The myth of mental illness: Foundations of a theory of personal conduct*. New York: Harper & Row.

Szyf, M. (2009). The early life environment and the epigenome. *Biochimica et Biophysica Acta, 1790* (9), 878–885.

Tauber, M. J., & Ackerman, D. (Eds.). (1990). *Mental models and human-computer interaction* (Vol. 2). Amsterdam: North-Holland.

Tennant, N. (1994). Changing the theory of theory change: Towards a computational approach. *British Journal for the Philosophy of Science, 45*, 865–897.

Tennant, N. (2003). Theory-contraction is NP-complete. *Logic journal of the IGPL, 11*, 675–693.

Tennant, N. (2006). New foundations for a relational theory of theory-revision. *Journal of Philosophical Logic, 35*, 489–528.

Thagard, P. (1988). *Computational philosophy of science*. Cambridge, MA: MIT Press.

Thagard, P. (1989). Explanatory coherence. *Behavioral and Brain Sciences, 12*, 435–467.

Thagard, P. (1992). *Conceptual revolutions*. Princeton: Princeton University Press.

Thagard, P. (1997). Coherent and creative conceptual combinations. In T. B. Ward, S. M. Smith, & J. Viad (Eds.), *Creative thought: An investigation of conceptual structures and processes* (pp. 129–141). Washington, DC: American Psychological Association.

Thagard, P. (1999). *How scientists explain disease*. Princeton, N.J.: Princeton University Press.

Thagard, P. (2000). *Coherence in thought and action*. Cambridge, MA: MIT Press.

Thagard, P. (2003). Pathways to biomedical discovery. *Philosophy of Science, 70*, 235–254.

Thagard, P. (2004). Causal inference in legal decision making: Explanatory coherence vs. Bayesian networks. *Applied Artificial Intelligence, 18*, 231–249.

Thagard, P. (2005a). *Mind: Introduction to cognitive science* (2nd ed.). Cambridge, MA: MIT Press.

Thagard, P. (2005b). The emotional coherence of religion. *Journal of Cognition and Culture, 5*, 58–74.

Thagard, P. (2006a). *Hot thought: Mechanisms and applications of emotional cognition.* Cambridge, MA: MIT Press.

Thagard, P. (2006b). How to collaborate: Procedural knowledge in the cooperative development of science. *Southern Journal of Philosophy, 44*, 177–196.

Thagard, P. (2006c). What is a medical theory? In R. Paton & L. A. McNamara (Eds.), *Multidisciplinary approaches to theory in medicine* (pp. 47–62). Amsterdam: Elsevier.

Thagard, P. (2007a). Abductive inference: From philosophical analysis to neural mechanisms. In A. Feeney & E. Heit (Eds.), *Inductive reasoning: Experimental, developmental, and computational approaches* (pp. 226–247). Cambridge: Cambridge University Press.

Thagard, P. (2007b). The moral psychology of conflicts of interest: Insights from affective neuroscience. *Journal of Applied Philosophy, 24*, 367–380.

Thagard, P. (2008a). Conceptual change in the history of science: Life, mind, and disease. In S. Vosniadou (Ed.), *International handbook of research on conceptual change* (pp. 374–387). London: Routledge.

Thagard, P. (2008b). How cognition meets emotion: Beliefs, desires, and feelings as neural activity. In G. Brun, U. Doguoglu, & D. Kuenzle (Eds.), *Epistemology and emotions* (pp. 167–184). Aldershot: Ashgate.

Thagard, P. (2008c). Mental illness from the perspective of theoretical neuroscience. *Perspectives in Biology and Medicine, 51*, 335–352.

Thagard, P. (2009). Why cognitive science needs philosophy and vice versa. *Topics in Cognitive Science, 1*, 237–254.

Thagard, P. (2010a). *The brain and the meaning of life.* Princeton, NJ: Princeton University Press.

Thagard, P. (2010b). EMPATHICA: A computer support system with visual representations for cognitive-affective mapping. In K. McGregor (Ed.), *Proceedings of the workshop on visual reasoning and representation* (pp. 79–81). Menlo Park, CA: AAAI Press.

Thagard, P. (2010c). Evolution, creation, and the philosophy of science. In R. Taylor & M. Ferrari (Eds.), *Epistemology and science education: Understanding the evolution vs. intelligent design controversy* (pp. 20–37). Milton Park: Routledge.

Thagard, P. (2010d). Explaining economic crises: Are there collective representations? *Episteme, 7*, 266–283.

Thagard, P. (2011). Critical thinking and informal logic: Neuropsychological perspectives. *Informal Logic, 31*, 152–170.

Thagard, P. (forthcoming-a). Coherence: The price is right. *Southern Journal of Philosophy.*

Thagard, P. (forthcoming-b). Conceptual change in cognitive science: The brain revolution. In W. J. Gonzalez (Ed.), *Conceptual revolutions: From cognitive science to medicine.* A Coruña, Spain: Netbiblo.

Thagard, P. (forthcoming-c). Mapping minds across cultures. In R. Sun (Ed.), *Grounding social sciences in cognitive sciences.* Cambridge, MA: MIT Press.

Thagard, P. (forthcoming-d). The self as a system of multilevel interacting mechanisms. Unpublished manuscript—University of Waterloo.

Thagard, P., & Aubie, B. (2008). Emotional consciousness: A neural model of how cognitive appraisal and somatic perception interact to produce qualitative experience. *Consciousness and Cognition, 17,* 811–834.

Thagard, P., & Croft, D. (1999). Scientific discovery and technological innovation: Ulcers, dinosaur extinction, and the programming langage Java. In L. Magnani, P. Nersessian, & P. Thagard (Eds.), *Model-based reasoning in scientific discovery* (pp. 125–137). New York: Plenum.

Thagard, P., & Litt, A. (2008). Models of scientific explanation. In R. Sun (Ed.), *The Cambridge handbook of computational psychology* (pp. 549–564). Cambridge: Cambridge University Press.

Thagard, P., & Shelley, C. P. (1997). Abductive reasoning: Logic, visual thinking, and coherence. In M. L. Dalla Chiara, K. Doets, D. Mundici, & J. van Benthem (Eds.), *Logic and Scientific Methods* (pp. 413–427). Dordrecht: Kluwer.

Thagard, P., & Stewart, T. C. (2011). The Aha! experience: Creativity through emergent binding in neural networks. *Cognitive Science, 35,* 1–33.

Thagard, P., & Verbeurgt, K. (1998). Coherence as constraint satisfaction. *Cognitive Science, 22,* 1–24.

Thompson, E. (2007). *Mind in life: Biology, phenomenology, and the science of mind.* Cambridge, MA: Harvard University Press.

Todorov, E., & Jordan, M. I. (2002). Optimal feedback control as a theory of motor coordination. *Nature Neuroscience, 5,* 1226–1235.

Treisman, A. (1996). The bindng problem. *Current Opinion in Neurobiology, 6,* 171–178.

Tripp, B. P., & Eliasmith, C. (2010). Population models of temporal differentiation. *Neural Computation, 22,* 621–659.

Tsankova, N., Renthal, W., Kumar, A., & Nestler, E. J. (2007). Epigenetic regulation in psychiatric disorders. *Nature Reviews. Neuroscience, 8*(5), 355–367.

Tsankova, N. M., Berton, O., Renthal, W., Kumar, A., Neve, R. L., & Nestler, E. J. (2006). Sustained hippocampal chromatin regulation in a mouse model of depression and antidepressant action. *Nature Neuroscience, 9*(4), 519–525.

Tweney, R. D., Doherty, M. E., & Mynatt, C. R. (Eds.). (1981). *On scientific thinking.* New York: Columbia University Press.

Unschuld, P. U. (1985). *Medicine in China: A history of ideas.* Berkeley: University of California Press.

Valdés-Pérez, R. E. (1995). Machine discovery in chemistry: New results. *Artificial Intelligence, 74*, 191–201.

van Rooij, I. (2008). The tractable cognition thesis. *Cognitive Science, 32*, 939–984.

von Neumann, J. (1958). *The computer and the brain.* New Haven: Yale University Press.

Vosniadou, S. (Ed.). (2008). *International handbook of research on conceptual change.* London: Routledge.

Vosniadou, S., & Brewer, W. F. (1992). Mental models of the earth: A study of conceptual change in childhood. *Cognitive Psychology, 24*, 535–585.

Wainfan, E., Dizik, M., Stender, M., & Christman, J. K. (1989). Rapid appearance of hypomethylated DNA in livers of rats fed cancer-promoting, methyl-deficient diets. *Cancer Research, 49*(15), 4094–4097.

Wall, L. (1996). *Programming Perl* (2nd ed.). Sebastopol, CA: O'Reilly.

Wang, D., Urisman, A., Liu, Y. T., Springer, M., Ksiazek, T. G., Erdman, D. D., et al. (2003). Viral discovery and sequence recovery using DNA microarrays. *PLoS Biology, 1*(2), E2.

Ward, T. B., Smith, S. M., & Vaid, J. (Eds.). (1997). *Creative thought: An investigation of conceptual structures and processes.* Washington, DC: American Psychological Association.

Warren, W. H. (2006). The dynamics of perception and action. *Psychological Review, 113*, 358–389.

Waterland, R. A., & Jirtle, R. L. (2003). Transposable elements: targets for early nutritional effects on epigenetic gene regulation. *Molecular and Cellular Biology, 23*(15), 5293–5300.

Weart, S. R. (2003). *The discovery of global warming.* Cambridge, MA: Cambridge University Press.

Weaver, I. C., Cervoni, N., Champagne, F. A., D'Alessio, A. C., Sharma, S., Seckl, J. R., et al. (2004). Epigenetic programming by maternal behavior. *Nature Neuroscience, 7*(8), 847–854.

Weisberg, R. W. (1993). *Creativity: Beyond the myth of genius.* New York: W. H. Freeman.

Welling, H. (2007). Four mental operations in creative cognition: The importance of abstraction. *Creativity Research Journal, 19,* 163–177.

Werning, M., & Maye, A. (2007). The cortical implementation of complex attribute and substance concepts: Synchrony, frames, and hierarchical binding. *Chaos and complexity letters, 2,* 435–452.

Wheeler, D. A. (2010). The most important software innovations. Retrieved Feb. 15, 2011, from http://www.dwheeler.com/innovation/innovation.html.

Whewell, W. [1840] (1967). *The philosophy of the inductive sciences.* New York: Johnson Reprint Corp.

Whewell, W. (1968). *William Whewell's theory of scientific method.* Pittsburgh: University of Pittsburgh Press.

Whorf, B. (1956). *Language, thought, and reality.* Cambridge, MA: MIT Press.

Wilson, E. O. (1998). *Consilience: The unity of knowledge.* New York: Vantage.

Wimsatt, W. C. (2007). *Re-engineering philosophy for limited beings.* Cambridge, MA: Harvard University Press.

Wisniewski, E. J. (1997). Conceptual combination: Possibilities and esthetics. In T. B. Ward, S. M. Smith, & J. Viad (Eds.), *Conceptual structures and processes: Emergence, discovery, and change* (pp. 51–81). Washington, DC: American Psychological Association.

Wittgenstein, L. (1968). *Philosophical investigations* (2nd ed., G. E. M. Anscombe, Trans.). Oxford: Blackwell.

Wolff, G. L., Kodell, R. L., Moore, S. R., & Cooney, C. A. (1998). Maternal epigenetics and methyl supplements affect agouti gene expression in Avy/a mice. *FASEB Journal, 12*(11), 949–957.

Wolpert, D. M., & Ghahramani, Z. (2000). Computational principles of movement neuroscience. *Nature Neuroscience, 3,* 1212–1217.

Woodward, J. (2004). *Making things happen: A theory of causal explanation.* Oxford: Oxford University Press.

Woodward, J. (2009). Scientific explanation. Retrieved February 11, 2011, from http://plato.stanford.edu/entries/scientific-explanation.

Wu, L. L., & Barsalou, L. W. (2009). Perceptual simulation in conceptual combination: Evidence from property generation. *Acta Psychologica, 132,* 173–189.

Yokomori, N., Moore, R., & Negishi, M. (1995). Sexually dimorphic DNA demethylation in the promoter of the Slp (sex-limited protein) gene in mouse liver. *Proceedings of the National Academy of Sciences of the United States of America, 92*(5), 1302–1306.

Young, J. O. (2008). The coherence theory of truth. *Stanford Encyclopedia of Philosophy.* Retrieved March 3, 2011, from http://plato.stanford.edu/entries/truth-coherence/.

Zhen, Z. (Ed.). (1997). *Zhong guo yi xue shi* (History of Chinese medicine; in Chinese) (2nd ed.). Shanghai: Shanghai Science and Technology Press.

Index